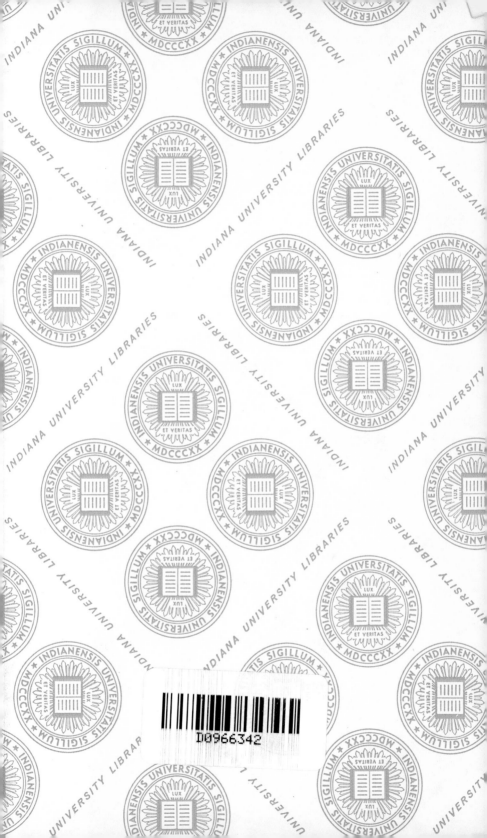

Correlation of the Silurian Rocks of the British Isles

By

A. M. ZIEGLER
Department of the Geophysical Sciences
University of Chicago
Chicago, Illinois 60637

R. B. RICKARDS
Sedgwick Museum
Cambridge, CB 3EQ
Great Britain

W. S. McKERROW
Department of Geology and Mineralogy
Oxford, OX1 3PR
Great Britain

Edited by
W.B.N. Berry and A. J. Boucot

THE GEOLOGICAL SOCIETY OF AMERICA

SPECIAL PAPER 154

Published by
THE GEOLOGICAL SOCIETY OF AMERICA, INC.
3300 Penrose Place
Boulder, Colorado 80301

Printed in The United States of America

*The printing of this volume has been made possible
through the bequest of Richard Alexander Fullerton Penrose, Jr.,
and the generous support of all contributors
to the publication program.*

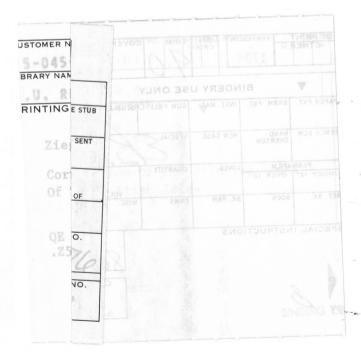

Contents

Acknowledgments

During the preparation of this paper, we consulted many British workers regarding their areas of special interest, and we are indebted to the following people for many helpful corrections and additions: R. J. Bailey, L.R.M. Cocks, J. R. Earp, R. R. Furness, D. C. Greig, C. H. Holland, J. D. Lawson, P. G. Llewellyn, A. R. MacGregor, D.J.W. Piper, J. F. Potter, W.D.I. Rolfe, J. H. Shergold, E. V. Tucker, V. G. Walmsley, E. K. Walton, P. T. Warren, J. A. Weir, D. E. White, H. E. Wilson, and A. Wood.

Financial support was provided by National Science Foundation Grant GP954X1 to the California Institute of Technology; Grants NSF GB6592 Res., GB17655 Res., and GA35073 to the University of Chicago; and by the Petroleum Research Fund Grant 910-G, a fund administered by the American Chemical Society.

Abstract

This paper is a compilation of published information considered pertinent to the correlation of the 450 Silurian stratigraphic units now recognized in the British Isles. The number of units reflects both the geologic complexity of these islands and the fact that much careful description and geologic mapping was done during the past two centuries. The accompanying bibliography contains more than 800 titles.

Graptolitic facies are widespread in the British Isles, and the graptolitic zonal sequence is used here to form the basis of correlation. In the shelf facies where graptolites are less common, brachiopod lineages are becoming increasingly useful in correlation. Recent work on other groups, particularly conodonts, spores, and ostracodes, provides further bases for correlation.

In early Paleozoic time, Britain and Ireland were traversed by an ocean basin, called the Iapetus Ocean, which in Silurian time, however, provided no barrier to migration as most marine animals are common to both sides. The ocean-floor deposits are characterized by thin, pelagic shale sequences and thick graywacke turbidite sequences, all within the graptolitic facies. Shelf deposits occur along both margins of the ocean, in the Welsh Borderland and Wales on the southeast, and in Scotland and Ireland on the northwest. These shelf areas are characterized by clastic shelly facies with limestone locally developed. Nonmarine deposits are mainly associated with Late Silurian time, just prior to the late phases of the Caledonian orogeny, when all of the British Isles except southwest England became a mountainous, nonmarine area.

1

Introduction

This paper compiles data on the correlation of the Silurian deposits of Great Britain and Ireland. It is an integral part of a series on Silurian correlation of the world, which is being compiled by W.B.N. Berry and A. J. Boucot. The portions on North America (1970), South America (1972a), Southeast Asia and the Near East (1972b), and Africa (1973) have been published (Geol. Soc. America Spec. Papers 102, 133, 137, and 147, respectively), and the preparation of similar sections for the rest of the world is nearly complete. The Silurian of the British Isles, the type region for the system, is complex and has been relatively well studied; therefore, a section devoted to this special area is warranted.

The initial part of the series on the North American Silurian contains an extensive introductory section, including material that is general to world-wide Silurian correlation, and also sections on such aspects of the North American Silurian as lithofacies relations, unconformities, and environmental relations. The reader is referred to this paper (Berry and Boucot, 1970) for some general correlation methods and to Ziegler (1970) for an interpretive synthesis of the development of the British geosynclinal complex during the Silurian period. However, some general information on correlation of particular relevance to the British Isles is included in the following introductory pages, as well as brief summaries concerning lithofacies, sediment sources, marine redbeds, communities, and unconformities.

The initial compilation of the material in this paper was made in 1966 by Ziegler, in consultation with Boucot and Berry, during the tenure of a postdoctoral fellowship at the California Institute of Technology. Rickards checked the text and made many changes and additions concerning the graptolite deposits. Later, the Geological Society of London published *A correlation of Silurian rocks in the British Isles* by Cocks and others (1971). That paper differs in scope and format from this study in that Cocks and others considered only the more important areas and that the information is presented by area rather than by stratigraphic unit. In this work, information is included that we consider relevant to the correlation of all stratigraphic units in current use, as well as an extensive bibliography of more than 800 titles. The authors of this work have been in consultation with Cocks and his colleagues during the preparation of both papers, and close agreement as to the correlation of most units has been achieved. During a late phase of this work, McKerrow helped to update and expand the original version. Capitalization and spelling of stratigraphic names in this paper are according to original British usage. Quotation marks have been inserted where usage subsequently has been found to be erroneous or misleading.

2

Correlation

The correlation standard adopted in the North American Silurian chart was the graptolite zonal sequence of Elles and Wood (1901-1918). It was thought that this sequence could be recognized on a world-wide basis and that subsequent charts in the series should conform to this standard. This zonal sequence is being refined by Rickards and his co-workers in Britain following earlier work, and both standard and refined schemes are illustrated (Fig. 1).

The graptolitic facies is widespread in the British Isles, and we place great reliance on graptolites for correlation purposes. In the nearshore areas, such as the Welsh Borderland where graptolites are rare, we have used brachiopods, but only the genera whose lineages are well understood. The stratigraphic distribution of brachiopods, like that of other benthic organisms, is controlled by ecologic as well as evolutionary factors, so that the range of such organisms in a section usually does not reflect its total duration in time. In the past, brachiopod assemblages have been used for correlation in an uncritical way, and serious mistakes have arisen. A good example of these mistakes is shown in the early studies of the Llandovery deposits of the Welsh Borderland. These rocks were deposited during a marine transgression, and each area shows the same sequence of assemblages. The early workers thought in terms of an "instantaneous transgression" at the base of the upper Llandovery, but Ziegler and others (1968b) showed that the assemblages were diachronous and that the transgression was extremely slow, having a duration of the whole of the Llandovery Age. In this study, brachiopods are used for correlation, but we used only those whose incremental evolutionary developments had been carefully worked out.

The graptolites and brachiopods are the main groups that have been used for correlation, but other groups are now being studied with this purpose in mind. Several papers have appeared recently on conodont occurrences in Britain (Aldridge, 1972; Austin and Bassett, 1967; Collinson and Druce, 1966; Brooks and Druce, 1965; Rhodes and Newall, 1963). The ostracode sequence of the Downtonian of the Welsh Borderland has recently been described and correlated with the sequences on the continent (Shaw, 1969). Finally, spore assemblages are now being described from the British Silurian (Richardson and Lister, 1969; Lister, 1970), and these should prove useful, particularly in correlating nearshore and nonmarine strata. The question of ecologic control of ostracodes and spores has yet to be critically examined. Such studies, as well as studies on their evolutionary relations, would enhance the value of these organisms for correlation.

LLANDOVERY SERIES

The Llandovery was a period of widespread graptolitic shale deposition, and the detailed correlation of these deposits was established in the late nineteenth and early twentieth centuries by Lapworth, Marr, Jones, and others. The stratigraphy of the shelf deposits of the Welsh Borderland was summarized recently by Ziegler and others (1968b), who used evolving trends in the brachiopods *Eocoelia*, *Stricklandia*, and *Pentamerus* as a basis for correlation. Correlation within the graptolitic and the shelly deposits is relatively routine, but there still are problems in correlating between these facies. Still another problem lies in relating these zones and lineages to the type area of Llandovery (cols. 50–52; these numbers refer to column numbers on Fig. 1).

The type area lies at the transition from shelf to basin, and although shelly faunas are present at many horizons, only three graptolite localities have ever been found. Of these, only one occurs in the central part of the Llandovery district where O. T. Jones (1925b) established the standard sequence (A_1-A_4, B_1-B_3, C_1-C_6). These tenuous ties are in the upper lower Llandovery (col. 51), the middle Llandovery (col. 50), and the basal upper Llandovery (col. 52). Until recently, there had been no evidence concerning the relation of graptolitic and shelly deposits during the uppermost Llandovery. However, several new finds, particularly in Shropshire and at Girvan, have been integrated in a new correlation system by Cocks (1971a), and this has necessitated a change in the standard as set out in the North American chart. The following stage names have recently been proposed for the Llandovery Series (Cocks and others, 1970): Rhuddanian for the lower Llandovery (A_1-A_4), Idwian for the middle Llandovery (B_1-B_3), Fronian for the lower upper Llandovery (C_1-C_3), and Telychian for the uppermost Llandovery (C_4-C_6).

WENLOCK SERIES

Relatively little attention has been devoted to the Wenlock until recently, and the correlation is based entirely on the graptolitic scale as worked out by Elles (1900) in the Builth area (col. 53). Recently, Bassett and others (1974) have re-examined the Wenlock of Wenlock Edge (col. 62). They have established for the first time that the early Wenlock zone of *Cyrtograptus centrifugus* is present, and they are proposing the stage name Sheinwoodian for the lower Wenlock from this zone to the *C. ellesae* zone and the stage name Homerian for the *C. lundgreni* and *Pristiograptus ludensis* zones at the top of the Wenlock.

LUDLOW SERIES

The relation of the graptolitic scale to the Wenlock-Ludlow boundary has recently been re-examined (Holland and others, 1969), and the base of the Ludlow is now taken as the zone of *Monograptus nilssoni*. Graptolites are rare in the Ludlow sequences in the inliers to the south and east; therefore, stratigraphers have relied on assemblages of benthic fossils and on lithologic criteria for correlation. We have generally accepted the correlations of Holland and others (1963), but we would not expect such correlations to be precisely time parallel in a sequence that is so obviously a regressive one. The assemblages are indicative of an environmental sequence and it would be unlikely that the same environments existed simultaneously throughout an area as large as Wales and the Welsh Borderland. Nevertheless, it is pointed out by one of us (Rickards) that the graptolite faunas that have been found parallel the units of Holland and others (1963). The Ludlow Series and the Eltonian, Bringewoodian, Leintwardinian, and Whitcliffian Stages were created by Holland and others (1963).

PRIDOLI SERIES

The top of the Ludlow coincides with a change to paralic and in turn fresh-water conditions; the correlation of the fish-bearing Downton with the marine sequences of the Continent is difficult. Allen and Tarlo (1963) placed the top of the Downton beneath the *Psammosteus* Limestone and the *Traquairaspis* zones, and we have taken the Downton, so defined, as broadly equivalent to the marine Pridoli Series of the Continent.

Distribution

Silurian rocks are widespread south of the Scottish Highlands and west of the English Midlands (Fig. 2). The Highlands are pre-Silurian metamorphic rocks, and much of England, with the exception of the Lake District and the Welsh Borderland, is covered by late Paleozoic and Mesozoic rocks. Silurian rocks appear as small inliers in the Midlands and have been penetrated by boreholes in southeast England. Unfortunately, drilling has usually stopped when lower Paleozoic rocks have been encountered, so that only a few meters of these beds are known from any given borehole. Claims have been made that Silurian rocks exist in Cornwall. These have been reviewed by Cocks and others (1971, p. 126–127), and the relevant papers are included in the bibliography of this work. Ordovician and Devonian fossils have been identified from the Meneage Breccias of this area, and some of the rocks previously thought to be Silurian have recently yielded Lower Devonian conodonts (House, in discussion of Lambert, 1965). Also, there is a recent claim of a Silurian age for upper Dalradian rocks in Banffshire, Scotland (Skevington, 1971). This is based on a re-examination of graptolites collected in the 1850s, but the lithology of the specimens is unlike that of the upper Dalradian, and the claim seems dubious on geological grounds (Ziegler, 1970).

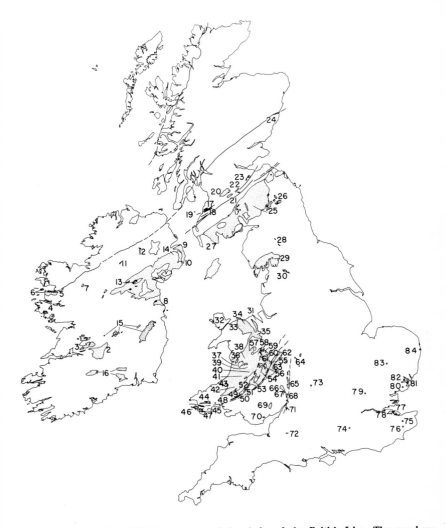

Figure 2. Distribution of Silurian rocks and boreholes of the British Isles. The numbers correspond to the columns on the correlation chart (Fig. 1). The stippled pattern represents the outcrop area, the circles are borehole locations, and the lines (dashed where approximately located) are faults known to have controlled sedimentation patterns during the Silurian.

Lithofacies Relations

A series of paleogeographic maps represen「ing five stages in the Silurian have been published (Ziegler, 1970) and are reproduced here (Figs. 3-7). The Silurian deposits of the British Isles are mainly geosynclinal in nature and trend across Ireland and Britain from southwest to northeast. The margins of this marine area can be seen in the Welsh Borderland on the southeast and in Scotland and the northern part of Ireland on the northwest. Each margin has nonmarine and shelf sediments and volcanic rocks associated with it as well as turbidites derived from it. Each margin has its own distinct history, and the only facies shared is the graptolitic shale facies that occupies the area between these margins.

NONMARINE SEDIMENTS

During Llandovery time, most of the land areas were undergoing erosion; therefore, nonmarine deposits are rare. There are thin sequences of redbeds at the base of marine transgressive sequences in the Malvern Hills, Herefordshire (col. 65), and northwest county Galway, Ireland (col. 4). The other known nonmarine bed of Llandovery age occurs interbedded with the Skomer Volcanic Group, Pembrokeshire (col. 46; Ziegler and others, 1969).

In the Wenlock, however, the marine basins along the northern margin, in the Midland Valley of Scotland (cols. 20-24), and in western Ireland (cols. 4-6) began to fill up, and the well-known Midland Valley fish faunas (Westoll, 1951) date from this time and may represent a paralic environment. The transition to nonmarine strata occurs somewhat later in Wales and the Welsh Borderland where, by definition, it coincides with the Ludlow-Downton (Pridoli) boundary. The assumption has always been made that the Ludlow Bone Bed, which marks the transition, is synchronous throughout this area, although this has never been proved. A considerable amount is known concerning the sequence of environments in the Downton. Judging from the sedimentary structures and abundant fish remains of these strata, Allen and Tarlo (1963) concluded that the Downton represents various paralic environments and that it is succeeded by fluviatile conditions in the Lower Devonian.

SHELF DEPOSITS

Shelf deposits are best developed on the southeast margin of the geosynclinal complex and extend from Pembrokeshire, in west Wales, through the Welsh

8

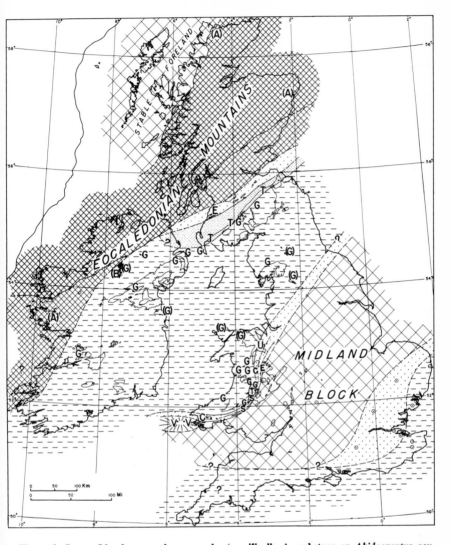

Figure 3. Lower Llandovery paleogeography (specifically A₂ substage or *Akidograptus acuminatus* and *Orthograptus vesiculosus* zones). Symbols for Figures 3 through 7 are as follows: land masses are cross-hatched in proportion to supposed relative elevation; shelf areas are represented by coarse dots; turbidite areas are fine dots; graptolitic shales and muds are horizontal dashes. A = age (each A represents two or more determinations falling within the 10-m.y. span suggested for the respective subdivisions of the Silurian); N = nonmarine strata; F = nonmarine strata containing fish remains; V = volcanic flow; L = *Lingula* Community; E = *Eocoelia* or *Cryptothyrella* Community; P = *Pentamerus* or *Pentameroides* Community; S = *Stricklandia* or *Costistricklandia* Community; C = *Clorinda* Community; U = undifferentiated shelly community; T = turbidite deposit; B = fluxoturbidite deposit; and G = graptolitic deposit. The parentheses indicate points of uncertain age or of age slightly different from that specified for the map. Arrows indicate paleocurrent directions as determined from sole structures of turbidites (from Ziegler, 1970).

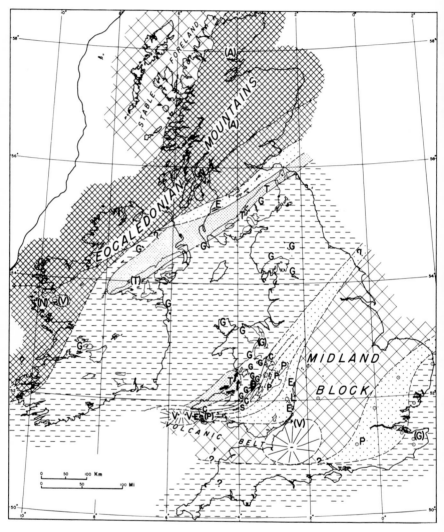

Figure 4. Early upper Llandovery paleogeography (specifically C_1-C_2 substages or *Monograptus sedgwicki* and *M. turriculatus* zones). For explanation of map symbols see Figure 2.

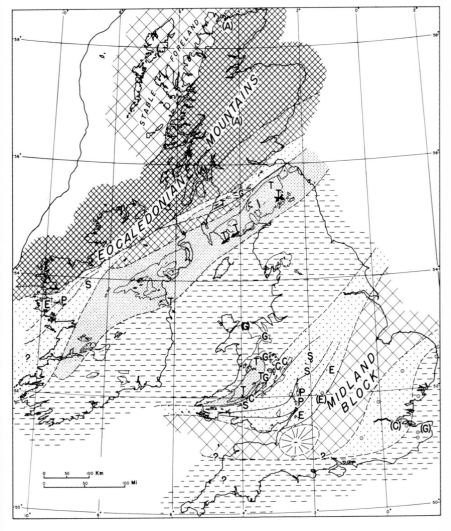

Figure 5. Latest Llandovery paleogeography (specifically C$_5$ substage or *Monograptus griestoniensis* zone). For explanation of map symbols see Figure 3.

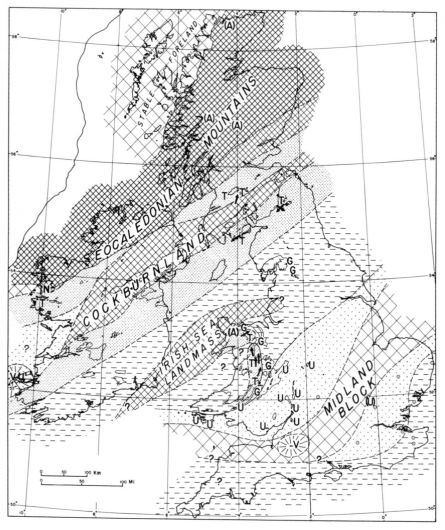

Figure 6. Early Wenlock paleogeography (specifically *Cyrtograptus murchisoni* and *Monograptus riccartonensis* zones). For explanation of map symbols see Figure 3.

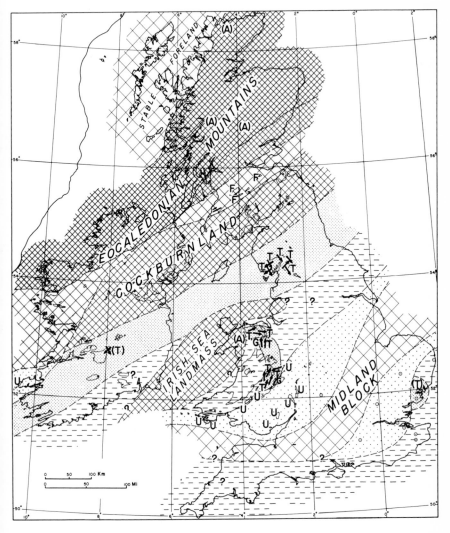

Figure 7. Early Ludlow paleogeography (specifically *Monograptus nilssoni* and *M. scanicus* Zones). For explanation of map symbols see Figure 3.

Borderland. These span the Llandovery, Wenlock, and Ludlow, and the shelf transgressed gradually toward the southeast during this time. Most of the deposits are clastic in nature, and these are extremely variable in composition, grain size, and thickness, reflecting local source areas and irregularities in the topography of the basement (Ziegler and others, 1968b). Carbonates, such as the Woolhope, Wenlock, and Aymestry Limestones, occur in the Wenlock and Ludlow.

The shelf fringing the northern margin is less well exposed and is seen only in three widely spaced localities: Girvan, Scotland (cols. 17-19); County Galway and surrounding areas (cols. 4, 5, 7); and Dingle, County Kerry (col. 1). The Girvan and Galway shelf sequences are of Llandovery age, while the Dingle sequence is Wenlock and Ludlow, although there the base is unknown.

TURBIDITES

The turbidites belong with the graptolitic facies in the (presumably) deeper water areas but do show a close relation to one shelf margin or the other. The Welsh turbidites occur in two areas, the Montgomery Trough of North Wales and the Denbigh Trough of central Wales (Cummins, 1969). Turbidite sedimentation began in the Montgomery Trough during late Llandovery time in west-central Wales (cols. 36-39, 42, 43), and the axis of maximum sedimentation moved eastward and eventually came to lie in east-central Wales in Wenlock and Ludlow time (cols. 53, 54, 56). These sediments seem to have been derived from the southern margin, but sediments in the Denbigh Trough came from the west and probably were derived from tectonic land on the site of the Irish Sea. Turbidite sedimentation in the Denbigh Trough began in the Wenlock and continued in the Ludlow (cols. 31, 33-35).

Turbidites derived from the Scottish side occur in the Southern Uplands Trough. These began to arrive prior to the Silurian, in Caradocian time, and show a continuous development throughout the Silurian. The Llandovery representatives of this great progression occur in the Southern Uplands (cols. 25-27) and its lateral extension in County Down, Ireland (col. 13). The Wenlock and Ludlow representatives extended farther south into the Lake District (cols. 28-30) and areas in central Ireland (cols. 2, 3).

GRAPTOLITIC SHALES

During the Llandovery, graptolitic shales were deposited over an extremely wide area; the best localities are near Moffat in the Southern Uplands (col. 25), in the Lake District (col. 29), and in central Wales (cols. 36-42). Recently, new localities have been discovered in Ireland (cols. 3, 13, 14), which make the distribution of this facies even more impressive. Rickards (1964b) has interpreted some black graptolitic shales as distal turbidites.

During the Wenlock and Ludlow, this widespread area of graptolitic shale deposition was progressively encroached on by proximal deposition. The best Wenlock localities are the Lake District (col. 29), Builth in central Wales (col.

53), and Long Mountain, Shropshire (col. 59). None of these localities, however, shows the typical attenuated graptolitic shale deposits of earlier times; the rocks are a mudstone facies that is transitional to the proximal turbidite faces. In the Ludlow, graptolitic deposits are harder to find; the only relatively continuous zonal sequences are located in the Lake District (col. 29), North Wales (cols. 31, 35), and Long Mountain (col. 59).

VOLCANIC ROCKS

Volcanic rocks in the British Silurian are limited to the margins of the geosyncline. Volcanic flow rock is known from three localities along the southern margin, that is, Marloes, Pembrokeshire (col. 46), Tortworth, Gloucestershire (col. 71), and the Eastern Mendip Hills, Somerset (col. 72). At Marloes, more than 500 meters of basic-to-acid volcanic rocks occur and are interbedded in their top horizons with shallow shelly beds of late Llandovery age. Only two lavas occur at Tortworth and these are probably late Llandovery. The section in the Eastern Mendips is incomplete, and all that can be said is that a few hundred meters of andesitic lava and tuff occur in association with shallow marine beds of probable Wenlock age. Bentonite beds are widespread in the Welsh Borderland and Wales. No fewer than 51 probable bentonite horizons were found in the uppermost Llandovery and Wenlock Beds of the Walsall borehole (col. 64; Butler, 1937, p. 254), and at least this many occur at the same horizons in the Howgill Fells.

Along the northern margin of the geosynclinal complex, volcanic rocks are known at Stonehaven, Scotland (col. 24); northwest County Galway (col. 4); and the Dingle Peninsula, County Kerry (col. 1). At Stonehaven, a few hundred meters of tuff and tuffaceous sandstone occur in beds associated with a Downton fish horizon. In Galway, a keratophyre lava up to 116 m thick occurs within the upper Llandovery Lough Mask Formation. Finally, the Wenlock and Ludlow beds of Dingle contain many rhyolite flows and ash beds.

Sources of Sediment

Ziegler (1970) has identified four major source areas in the Silurian of the British Isles. The Welsh Borderland and adjacent English Midlands provided both contemporaneous volcanic debris and the products of erosion of a great variety of earlier rocks to the Welsh Basin. This was probably the major source of sediment for the basin, although there is evidence for tectonic land on the site of the Irish Sea in Wenlock and Ludlow time. Turbidites in the Denbigh Trough of North Wales could only have come from this direction, and the massive slumping of these beds is a further indication of tectonic instability.

Along the northern margin of the geosynclinal belt, the turbidites of the Southern Uplands are composed mainly of basic volcanic rocks of the type now exposed along the margins of the Midland Valley. Highland metamorphic rocks also occur, indicating that much of Scotland must have been emergent during the Silurian. This is the Eocaledonian source of Wills (1951), and there is evidence of still another land area, which is called Cockburnland by Walton (1963). Cockburnland was tectonic in nature and shed sediments north, into the Midland Valley, and south, toward the Lake District, during Wenlock and Ludlow time. It consisted of folded graywackes of Ordovician and probably Llandovery age, so that in a real sense the geosyncline was feeding on itself.

Marine Redbeds

Red shale and mudstone deposited in quiet, offshore marine environments are a striking feature, particularly of the upper Llandovery sequences of many parts of the British Isles, and of sequences in Norway, Estonia, and the Appalachian Basin as well. Ziegler and McKerrow are preparing a manuscript on the distribution and origin of these interesting sediments. In the British Isles, they occur in Idwian or Fronian sequences in Scotland (col. 25) and Wales (cols. 35, 40, 44); in lower Telychian sequences in Scotland (cols. 18, 25, 27), Wales (cols. 35, 44, 57), and the Welsh Borderland (cols. 59, 60, 62, 64); in upper Telychian sequences of Ireland (cols. 4, 8), Scotland (cols. 18, 23, 25-27), northern England (cols. 28, 29), Wales (cols. 38, 39, 42, 53), and the Welsh Borderland (cols. 62, 64-66); and finally in the lower Wenlock sequence of Scotland (cols. 26, 27).

Ziegler and McKerrow (unpub. data) found that the red shales were confined to the offshore shelf and deeper marine areas (*Stricklandia* and *Clorinda* Communities and graptolitic facies), and that within these environments they are found only where rates of sedimentation were most rapid. Their occurrence is correlated with transgressive pulses; therefore, they probably resulted from relatively rapid coastal erosion of thoroughly weathered terrain. Apparently, their preservation in the oxidized state was possible only in quiet, offshore areas where rates of sedimentation were high. They are not universally distributed during any particular interval, and they range in age from middle Llandovery to early Wenlock; hence, they are not particularly useful for correlation purposes.

Communities

Silurian marine communities were initially defined from the upper Llandovery beds of the Welsh Borderland (Ziegler, 1965; Ziegler and others, 1968a). These communities were named for characteristic genera within them, for example, *Lingula*, *Eocoelia*, *Pentamerus*, *Stricklandia*, and *Clorinda*, and they represent progressively offshore shelf conditions. Several of the characteristic genera range through only part of the Silurian, but the communities usually can be recognized on the basis of the other elements. Thus, *Eocoelia* has a limited range of upper Llandovery and lower Wenlock, but elements of the *Eocoelia* Community are associated with *Cryptothyrella* in lower and middle Llandovery rocks, so here the word *Cryptothyrella* Community may be used.

In Wenlock and Ludlow rocks, the characteristic genera, with the exception of *Lingula*, are absent, but Cummins (1969) has traced "probable depth indicators" of the Llandovery into these later strata. He used *Salopina* as an indicator of the *Eocoelia* Community and found it most abundant at the top of the Ludlow in Leintwardinian and particularly Whitcliffian beds. This is the herald of the regression at the top of the Ludlow. Cummins believed that *Atrypa* and *Eospirifer* indicated the *Pentamerus* and *Stricklandia* Communities, and he found these to be particularly abundant in both the Wenlock Limestone and middle Ludlow beds. Finally, the associates of the *Clorinda* Community—*Dicoelosia*, *Cyrtia*, and *Skenidioides*—occur in the Wenlock Shale and lower Ludlow Shale (Elton Beds). This conforms to the expectation that the shale-limestone alternations of the Welsh Borderland correspond to quiet, offshore conditions and to shallower conditions, respectively, and that the beds at the top of the Ludlow represent still shallower water. The details of the Wenlock and Ludlow communities are currently being worked on in a series of projects for theses at Oxford University by Calef, Hancock, and Hurst (1973, personal commun.). They recognize four communities, *Salopina*, *Sphaerirhynclia*, *Isorthis*, and *Dicoelosia*, which correspond broadly with the *Eocoelia*, *Pentamerus*, *Stricklandia*, and *Clorinda* Communities, respectively.

The Downton, which is above the Ludlow, represents paralic conditions with abundant fish remains as well as *Lingula*, some bivalves, gastropods, and ostracodes (Allen and Tarlo, 1963). This community was eventually replaced by a purely fluviatile fish fauna in beds of the higher Ditton.

Unconformities

In the Welsh Borderland and adjacent parts of Wales, a major unconformity separates Silurian strata from rocks of Ordovician and pre-Ordovician age (cols. 57-71). This represents a major withdrawal of the sea near the end of the Ordovician and has been related to the Taconic orogeny by some authors. However, the angular discrepancy of beds of Silurian and Ordovician age is not great, and there is little evidence of substantial compression (Ziegler, 1970). A more important effect may have been a eustatic change at about the end of the Ordovician, due to glaciation in Africa. Also, in the Welsh Borderland many disconformities have been detected in the shelf sequences (cols. 65-68). These may be related to periods of nondeposition or erosion in the shallow-water areas. Along the shelf margin in Wales angular unconformities have been detected, and these appear to be related to tectonic instability along the shelf-basin hinge (cols. 44-54).

Along the northern margin of the marine area in the British Isles, there is an unconformity between beds of Silurian and Ordovician age in Scotland (cols. 17-19) and in western Ireland (cols. 4, 5, 7). In the beds of Wenlock and Ludlow age of the Midland Valley (cols. 20-23), gaps are thought to occur, and these may be early manifestations of the major folding that occurred in this and other areas toward the end of the Silurian. This folding is evident in Ireland (cols. 1-3, 7, 11, 12), Scotland (cols. 20-23), and Wales (cols. 47, 48), although it is absent from the more stable Welsh Borderland. It is difficult to determine the time span of this folding because the angular unconformity is succeeded everywhere by nonmarine strata. Pteraspid fish remains occur in many areas above this unconformity, so that presumably the folding occurred in latest Silurian or earliest Devonian time.

Alphabetical List of Rock Units

The numbers in parentheses refer to column numbers on the correlation chart (Fig. 1) and to corresponding numbers on the accompanying map (Fig. 2); see the introductory section of the bibliography for a grouping of the column numbers referring to geographic areas.

Note that no attempt has been made to change stratigraphic terminology or fossil names from those used in the last quoted paper on an area.

A_1-A_4, A_a-A_c
(50, 51, 52, 53)

See lower Llandovery of Llandovery and Garth.

ABERYSTWYTH GRITS
(43)

Jones (in Wood and Smith, 1959, p. 191) has stated that the deposition of the Aberystwyth Grits of the Llangranog area began in the lower part of the zone of *Monograptus turriculatus.* The upper limit of this series is not seen in the vicinity. The equivalent unit to the northeast is the Ystwyth Stage (cols. 36, 37), which is often referred to as the Aberystwyth Grits.

ACUMINATUS SHALES
(39)

See Clywedog "Stage."

AUSTWICK FLAGS AND GRITS
(30)

King and Wilcockson (1934, p. 21) have identified the zones of *Cyrtograptus murchisoni* and *Monograptus riccartonensis* in this unit. The specimen of *C. rigidus* referred to by Furness and others (1967, p. 138) came from this unit and indicates the presence of the *C. rigidus* zone.

AYMESTRY FORMATION OF THE MALVERN DISTRICT
(65)

Phipps and Reeve (1967, p. 347-349) have recently subdivided this unit into a lower Rilbury Siltstone and an upper Aymestry Limestone. Graptolites have yet to be found in the Malvern area; therefore, the only resort is to correlate this formation with the Bringewoodian Stage of Ludlow, which it resembles in general lithologic and faunal aspects. In addition, the upper part of the Aymestry Limestone in the Malvern area contains a faunal assemblage typical of the Lower Leintwardine Beds (Mocktree Shale), and Phipps and Reeve believed that this part of their unit was time correlative with the Lower Leintwardine Beds of Ludlow.

AYMESTRY LIMESTONE OF THE MALVERN DISTRICT
(65)

See Aymestry Formation of the Malvern district.

B, B_1-B_3, B_a-B_d
(51, 52, 53)

See middle Llandovery of Llandovery and Garth.

BAILEY HILL BEDS
(54)

Holland (1959b, p. 453) has reported *Monograptus bohemicus* from the lower horizons of the Bailey Hill Beds, indicating a correlation with the *M. nilssoni* or *M. scanicus* zones, and *M. leintwardinensis* from the higher horizons, establishing the presence of that zone. Intermediate horizons have yielded *M. chimaera* var. *semispinosus* and *M. leintwardinensis* var. *incipiens*.

"BALA" OF LLANDOVERY, GARTH, AND BUILTH
(50, 51, 52, 53)

Andrew (1925, p. 391) collected *Glyptograptus* cf. *persculptus* and *Mesograptus* cf. *modestus* var. *parvulus* from the top of the "Bala" rocks at Garth; evidently they span the Ordovician-Silurian boundary.

BANNISDALE SLATES
(29)

Furness (1965, p. 252) has stated that the lower part of the Bannisdale Slates contains the zone of *Monograptus leintwardinensis incipiens*. Rickards (1965, p.

248) would correlate this zone with the *M. tumescens* zone of Elles and Wood. Rickards (1967) in addition, has collected a *M. nilssoni-M. scanicus* assemblage and a *M. leintwardinensis* assemblage, all from the lower part of this unit.

BARR LIMESTONE
(64)

The Barr Limestone is a relatively thin unit that rests just above beds containing *Cyrtograptus murchisoni;* therefore, this limestone must occupy a position slightly above the base of the Wenlock (Butler, 1937, p. 243, 246). It is overlain by the Wenlock Shales, which have yielded only *Monograptus priodon,* a species which has an upper limit of the zone of *M. riccartonensis* or possibly that of *Cyrtograptus rigidus.*

BASAL BEDS
(29)

See Stockdale Shales.

BEDS E TO H OF THE NORTH ESK INLIER
(23)

These beds are mostly mudstone and have yielded a single graptolite identified by Bulman as *Monograptus vomerinus* aff. var. *gracilis* (Mitchell and Mykura, 1962, p. 21). There is some doubt about the validity of the subspecies *gracilis,* and slender monoclimacids of this type can only be taken to indicate a late Llandovery to early Wenlock age.

BENARTH FLAGS AND GRITS
(33)

The zone of *Cyrtograptus murchisoni* is well developed in the basal beds of this unit (Elles, 1909, p. 186-187). Above this are beds with *M. riccartonensis* as the most abundant fossil, and it is within these beds, which probably represent the zone of *M. riccartonensis,* that the first massive grits appear. These grits are known elsewhere as the Denbigh Grits (Cummins, 1957, p. 434). Elles (1909, p. 187) believed that they extended into the overlying zone of *C. symmetricus* (=*C. rigidus*) because of the presence of *M. dubius.* However, it is now known that *M. dubius* commonly occurs much earlier, and its presence in abundance here could also indicate a high *M. riccartonensis* zone. The top of the unit is not known in the area that Elles mapped, but higher beds occur to the southeast (see col. 34).

BIRKENHEAD SANDSTONE
(20)

See Waterhead Group.

BIRKHILL SHALES
(25)

The Birkhill Shales constitute the upper unit of C. Lapworth's Moffat Series (1878b, p. 250), one of the standard sequences for graptolite zonation in the Upper Ordovician and Lower Silurian. Lapworth included three zones from the base in the Lower Birkhill Shales: the zones of *Akidograptus acuminatus, Cystograptus vesiculosus,* and *Monograptus gregarius.* Recently, Toghill (1968b) has also recognized the zones of *Glyptograptus persculptus* and *Monograptus cyphus.*

In the Upper Birkhill Shales, Lapworth recognized the zones of *Cephalograptus cometa, Monograptus sedgwicki,* and *Rastrites maximus.* Elles and Wood (1901-1918) recognized the zone of *Cephalograptus cometa* as a band in the zone of *Monograptus convolutus,* and the zone of *Rastrites maximus* was recognized as a band at the base of the zone of *Monograptus turriculatus.* Graywackes of the Gala Group directly overlie Lapworth's zone of *Rastrites maximus.*

BLACK COCK BEDS
(48, 49, 50, 51, 52)

The age of the Black Cock Beds is uncertain. Potter and Price (1965, p. 397, 399) regarded the shelly fauna as broadly similar to the Lower Bringewood Beds of the type area and argued that the comparatively great stratigraphic thickness of the Black Cock Beds would suggest that they probably include other horizons as well, for example, the Middle and Upper Elton Beds. They collected only one graptolite species (Potter and Price, 1965, p. 393), *Monograptus* cf. *M. colonus,* and this would support a correlation with the Middle Elton Beds.

Potter and Price recognized local facies at the very bottom and top of the Black Cock Beds, which they termed the Lletty Bed Facies and the Carn Powell Facies, respectively. The Lletty Bed Facies is developed only in the Llandeilo area, whereas the Carn Powell Facies is developed in the Llangadock area and probably also in the Llandeilo area.

BLACK SHALES AND FLAGS OF PORTAFERRY
(10)

Swanston and Lapworth (1878, Table 1) gave an apparently composite list of graptolites from old quarries at Tieveshilly near Portaferry. Most of the species are common to the zones of *Monograptus turriculatus* and *M. crispus,* and in fact, both of these species occur. In addition, *M. cyphus* and *M. riccartonensis* occur; they are zonal indices for lower Llandovery and lower Wenlock zones, respectively. Swanston and Lapworth (1878, p. 121) stated that all the fossils "undoubtedly occupy only a black slaty band," so that, assuming the identifications are correct, the sequence must be a condensed one.

BLACK SHALES OF DONAGHADEE
(9)

Lapworth identified and described many graptolites from this area (Swanston and Lapworth, 1878). He recognized all the Llandovery graptolite zones up to the base of the *Rastrites maximus* subzone. The biostratigraphy of the area is being revised by members of The Geological Survey of Northern Ireland; they have defined the zones of "*M. cyphus, M. gregarius,* and the lower part of *M. convolutus*" (H. E. Wilson, 1967, written commun.). They have not been able to confirm the *A. acuminatus* and *O. vesiculosus* zones recognized by Lapworth.

BLACK SHALES OF SAINTFIELD
(14)

Small exposures in this district have yielded graptolite faunas in the *Monograptus gregarius* zone (Pollock and Wilson, 1961) and in the *Akidograptus acuminatus* and (?) *Cystograptus vesiculosus* zones (H. E. Wilson, 1967, written commun.,). The beds are in apparently conformable succession with the uppermost zones of the Ordovician.

BLACK SLUMPED MUDSTONES OF HARWICH
(81)

The original description of this borehole is to be found in Prestwich (1858, p. 250-252), who reported that the drill penetrated 13.5 meters beneath the basal Gault horizon. The reported specimen of *Posidonia* has been shown to be inorganic (Bullard and others, 1940, p. 87), and in the absence of fossil evidence, the rocks are correlated with the nearby Stutton borehole beds of early Wenlock or late Llandovery age.

BLAEBERRY FORMATION
(20)

See Priesthill Group.

BLAIR SHALES
(18)

This unit contains the *M. crenulata* zone (Cocks and Toghill, 1973).

BLUE-BLACK SHALES
(40, 41)

These Upper Ordovician beds (H. Lapworth, 1900, p. 75) have recently yielded the *Glyptograptus persculptus* zonal assemblage (Kelling and Woolands, 1969, p. 259), showing that they may extend into the lowest Llandovery.

BOG QUARTZITE
(61)

The term "Bog Quartzite" refers to some loose boulders of the Shelve area that have yielded an abundant shelly fauna. Whittard (1932, p. 380) believed the Bog Quartzite to be of relatively uppermost Llandovery age, that is, equivalent to the Purple Shales (Hughley Shales), but Ziegler and others (1968b, p. 743) have reported *Stricklandia lens intermedia* from these boulders, which would suggest a middle Llandovery age.

BRATHAY FLAGS
(28, 29)

See Lower Coniston Flags.

BRINGEWOOD-ELTON BEDS OF TORTWORTH
(71)

These mudstones and thin limestones from the Brookend borehole were correlated with the Bringewood and Elton Beds of Ludlow (Cave and White, 1968), presumably on general lithologic and faunal similarity.

BRINKMARSH BEDS
(71)

This name has been proposed (Curtis, 1972, p. 18) for beds at Tortworth that may be correlated with the Wenlock of areas to the north because of vague lithologic and faunal similarities. They are conformable above uppermost Llandovery beds, and their contact with the Ludlow is obscured by faulting. Ziegler (1966, p. 538) has reported *Eocoelia sulcata* from these beds.

BROADFORD GROUP
(3)

The Broadford Group contains, at one horizon, a large brachiopod fauna (Weir, 1962, p. 249), including *Eospirifer* sp. and *Cyrtia* sp., which establishes a lower limit of C_3, and *Pentamerus,* which has not been found in the British Isles above C_5. The overlying Craglea Group has yielded hooked monograptids that Weir (1962) believes indicate a late Wenlock age; on this evidence, the age could in fact be late Llandovery to Ludlow.

BROWGILL BEDS
(28, 29, 30)

See Stockdale Shales.

BRYNMAIR GROUP
(38)

See Tarannon "Series."

BRYN-Y-SAESON BEDS
(31)

A *Monograptus leintwardinensis* zonal assemblage was reported from these beds by Woods and Crosfield (1925, p. 180). The top of the unit is unknown; it is cut out by the Carboniferous angular unconformity.

BUILDWAS BEDS
(62)

See Wenlock Shales.

BURN BRIDGE ARENITE
(22)

See Carmichael Burn Group.

BUTTINGTON MUDSTONE FORMATION
(59)

This unit was described by Wade (1911, p. 436), and it was also mentioned by Whittard (1932, p. 881-882) who referred to it as the Purple Shales. Palmer (1970a, p. 342) proposed the present name. The fossils discovered in the Buttington Mudstone Formation are not diagnostic of age, but the color suggests a correlation with the latest Llandovery Hughley Shales (Purple Shales) of Shropshire.

BWLCH-PEN-BARRAS BEDS
(31)

Woods and Crosfield (1925, p. 174-175) collected graptolites from this unit that are indicative of the zone of *Monograptus nilssoni.*

C_1-C_6, C_a-C_d
(50, 51, 52, 53)

See upper Llandovery of Llandovery, Llangadock, and Garth.

CABAN GROUP
(41)

H. Lapworth (1900) initially defined this group and regarded it as stratigraphically above the Gwastaden Group of the Rhayader area. Kelling and Woolands (1969), however, reinterpreted the stratigraphy of these areas and concluded that these two groups are facies of each other. They include the Gerig Gwynion Grits, the Dyffryn Flags, the Caban Conglomerate with its intermediate shales, and the Gafallt Beds in the Caban Group. The *A. acuminatus* zone was reported from the lower Dyffryn Flags in the adjacent Rhayader area (H. Lapworth, 1900, p. 76), and a meager fauna from the top of this unit would indicate that it extends at least to the *triangulatus* zone (Kelling and Woolands, 1969, p. 259). The intermediate shales of the Caban Conglomerate have yielded graptolites indicating either *triangulatus* or *magnus* zones (Kelling and Woolands, 1969), whereas the Gafallt Beds contain the *sedgwicki* zone (H. Lapworth, 1900, p. 101).

CADDROUN BURN BEDS
(26)

See Riccarton Group.

CAERAU GROUP
(39)

See Llanidloes "Stage."

CANASTON BEDS
(44)

See Millin "Stage."

CAPEL BERACH BEDS
(48)

This unit, like the underlying Cwm Lane Beds, has not yielded a diagnostic fauna (Potter and Price, 1965, p. 396). The beds are the probable equivalent of part of the Red Marls farther to the northeast and have been regarded as Downton in age.

CARGHIDOWN BEDS
(27)

See Hawick Rocks.

CARMICHAEL BURN GROUP
(22)

Three units constitute the Carmichael Burn Group: the Crossridge Formation, the Burn Bridge Arenite, and the Manse Mudstones (Rolfe, 1960, p. 247-251). The lowest, the Crossridge Formation, has yielded *Monograptus vomerinus* var. *gracilis, M.* cf. *griestoniensis,* and *Plegmatograptus obesus* cf. var. *macilentus,* which Strachan (*in* Rolfe, 1960) thought indicated a late Llandovery age. The base of the Carmichael Burn Group is unknown.

CARN POWELL FACIES
(48, 49)

See Black Cock Beds.

CARRICKALLEN PSAMMITIC FORMATION
(13)

This unfossiliferous unit rests with probable erosional unconformity on the Lough Acanon Pelitic Formation of Late Ordovician and earliest Llandovery age (Phillips and Skevington, 1968, p. 143).

CARRIGHILL FORMATION
(15)

See Shales, Grits, and Conglomerates of Slieve Bloom, Knockshigowna, Blessington, and Chair of Kildare.

CARTLETT BEDS
(44)

See Haverford "Stage."

CASTELL GROUP
(37)

See Pont Erwyd "Stage."

CASTLE FORMATION
(20)

See Priesthill Group.

CAUSEMOUNTAIN FORMATION
(59)

Palmer (1970a) named this unit and correlated it with the Whitcliffian and basal Downton (=Pridoli), apparently because of general faunal and lithologic similarities.

CEFN FORMATION
(59)

Wade (1911, p. 435) described this unit, and it was later redescribed by Whittard (1932, p. 881) as the *Pentamerus* Beds in conformity with his Shropshire stratigraphy. The present name was proposed by Palmer (1970a, p. 342). *Stricklandia lens* cf. *intermedia* has been reported from the Cefn Group by Ziegler and others (1968b, p. 766), and this would indicate a middle Llandovery age. The Cefn Formation rests unconformably on the Gwern-y-brain Group of latest Ordovician age at Cefn, but to the northeast it rests on successively lower horizons of the Ordovician.

CEFN GOLEN BEDS
(31)

Woods and Crosfield (1925, p. 179) reported a *Monograptus tumescens* faunal assemblage from these beds. For the purposes of the present correlation chart, we include this assemblage within the *M. scanicus* zone.

CERIG GWYNION GRITS
(40, 41)

See Gwastaden Group and Caban Group.

CHONETES BEDS OF BUCKNELL
(55)

Stamp (1919, p. 228-231) described these beds, which occur directly below the Downton Castle Sandstone. They have not yielded a diagnostic fauna (graptolites or brachiopods that are presently useful), and Holland and others (1963, Table II) assigned them to the upper Whitcliffian.

CLIFFORD'S MESNE BEDS
(67, 68)

Lawson (1954, p. 232-233; 1955, p. 107-108) described these beds and compared them with the Grey Downton Formation of the Ludlow.

CLODDIAU GROUP
(58)

Wade (1911, p. 434) reported *Climacograptus innotatus, C.* cf. *rectangularis,* and *C. medius* from the lower beds of this group; these three forms are known to range throughout the lower Llandovery and into the middle Llandovery, that is, from the *Akidograptus acuminatus* zone to the *Monograptus gregarius* zone. Unfortunately, the upper limit of the Cloddiau Group cannot be established at this time, but the report of *Pentamerus oblongus* indicates that middle or upper Llandovery beds are included.

CLONCANNON BEDS
(2)

These beds contain a graptolite fauna that was originally interpreted to indicate the zone of *Monograptus tumescens* (Cope, 1959, p. 223). This fauna has recently been re-examined (Palmer, 1970b) and was reported to contain *M. ludensis* and *M. auctus,* indicating the upper Wenlock zone of *M. ludensis.* Lower Paleozoic strata of the Devilsbit Mountain district are overlain by the Old Red Sandstone with marked angular unconformity.

CLOONTRA GROUP
(3)

The Cloontra Group (Weir, 1962) is unfossiliferous, but a late Llandovery age is indicated by its probable stratigraphic position above graptolitic shale that contains a *M. turriculatus*-zone fauna and below the Broadford Group that contains latest Llandovery brachiopods.

CLOUGHER HEAD FORMATION
(1)

See Dunquin Group.

CLYWEDOG "STAGE"
(39)

Five units have been recognized within the Clywedog "Stage": the *Persculptus* Mudstones, the *Acuminatus* Shales, the *Monograptus* Beds, the *Triangulatus* Beds, and the *Convolutus* Beds (W.D.V. Jones, 1945). The basal unit, the *Persculptus* Mudstones, contained what Jones called the zone of *Glyptograptus persculptus*, a zone recognized throughout central Wales where it is usually taken as the base of the Llandovery. The *Acuminatus* Shales contain the zone of *Akidograptus acuminatus*. Jones recognized three zones in the *Monograptus* Beds, the zones of *M. atavus*, *M. acinaces*, and *M. cyphus*. The two top units of the Clywedog Stage, the *Triangulatus* Beds and the *Convolutus* Beds, are equivalent to the zones of *M. gregarius* and *M. convolutus*, respectively.

COALBROOKDALE BEDS
(62)

See Wenlock Shales.

CONISTON GRITS
(28, 29)

The Upper and Lower Coniston Grits are separated by a thin unit, the Sheerbate Flags (Furness and others, 1967, p. 136). Blackie (1933, p. 94, 100) has reported *Monograptus colonus*, *M.* cf. *roemeri*, and *M.* cf. *chimaera* var. *salweyi* from the base of the Coniston Grits. These species suggest a correlation with the zone of *M. nilssoni*. The main thickness of the Coniston Grits is poorly fossiliferous, but the overlying unit, the Bannisdale Slates, contains the zone of *Monograptus leintwardinensis incipiens* according to Furness (1965, p. 252), and this zone is presently correlated with the zone of *M. scanicus*.

The Coniston Grits have been mapped in the Cross Fell inlier (Nicholson and Marr, 1891; Shotton, 1935) but no fossils have been reported. Apparently, the identification of the unit in this area was based on lithologic similarities with the Lake District.

CONVOLUTUS BEDS
(39)

See Clywedog "Stage."

CORALLIFEROUS "SERIES"
(45, 46)

These beds rest with a slight angular unconformity on the Skomer Volcanic Group (Ziegler and others, 1969, p. 430, 434). This unconformity is visible now, both at Renney Slip and at Marloes Bay; the beds of the Skomer Volcanic Group were tilted about 5 or 10 degrees and reddened to a depth of 1 meter before the Coralliferous "Series" was deposited. The lower beds of the Coralliferous "Series" contain *Costistricklandia lirata*, and *Palaeocyclus* sp. occurs in the lower two-thirds of the unit; both fossils indicate a C_6 age. There are no diagnostic faunal elements in the upper third of the unit; presumably it extends into the Wenlock.

Similar beds occur in the Winsle anticline, but they have not yielded *Palaeocyclus* and their base is not seen. This area is currently being studied by students at Oxford University, who report a gradational contact with the Red Marls. Thus, these beds may include horizons that are time equivalents of the Sandstone "Series" of Marloes.

CORRYCROAR GROUP
(12)

The only fossils that have been found in this group are "two fragmentary graptolites with cells of the type of *M. vomerinus*" (Fearnsides and others, 1907, p. 109). Unfortunately, the Corrycroar Group is not exposed in the area where the underlying Little River Group occurs but "boulders probably referable to them, and not far travelled, are abundant in the Lime Hill district adjoining the exposure yielding *M. sedgwickii* [sic]" (Fearnsides and others, p. 110). Probably, then, the Corrycroar Group follows directly the Little River Group.

COWLEIGH PARK BEDS
(65)

These beds rest with slight angular unconformity on Cambrian shales west and northwest of the Malvern Hills. The highest beds have yielded *Eocoelia hemisphaerica* from the Old Storridge Common area, indicating an age of C_1 or C_2 (Ziegler and others, 1968b, p. 752). The Cowleigh Park Beds are overlain unconformably by the Wych Beds of C_5 to C_6 age, and in fact, the time gap represented by the unconformity can be accurately determined because *E. hemisphaerica* occurs in the beds immediately beneath the unconformity and *E. curtisi* occurs just above it.

CRAGLEA GROUP
(3)

A few graptolites have been found in the Craglea Group of Slieve Bernagh, including *Monograptus flemingi* and *M.* aff. *vicinus* from one horizon, and *M. bohemicus* from a higher horizon (Weir, 1962, p. 250) indicating a late Wenlock to early Ludlow age.

The nearby Slieve Aughty area has not been studied in recent years, but Harper (1948, p. 59) has summarized the early work of the Geological Survey of Great Britain. He mentioned the record of *Graptolithus priodon* but pointed out that

Survey identifications were poor indeed, and he concluded that the lithologies represented are commensurate with a Silurian correlation.

CRAIGSKELLY CONGLOMERATE
(19)

This conglomerate lies directly below the Woodland Formation with *M. cyphus* zone graptolites, so it is presumably of early Llandovery age. C. Lapworth (1882, p. 641) listed *Atrypa hemisphaerica* from Craigskelly, but Reed (1917, p. 955) rightly doubted Lapworth's identification of this species.

CREGGANBAUN GROUP
(5)

See Croagh Patrick "Series."

CROAGHMARHIN FORMATION
(1)

See Dunquin Group.

CROAGH PATRICK "SERIES"
(5)

These rocks have been described by three authors in recent years (Anderson, 1960a; Stanton, 1960; Dewey, 1963). Each author has used a separate terminology, and there has been no attempt to relate one set of stratigraphic terms to another; therefore, the situation is confusing. We are adopting Dewey's nomenclature and are indebted to D.J.W. Piper (1967, written commun.) for helping us relate Dewey's terms to earlier publications.

Dewey's Croagh Patrick "Series" consists, from bottom to top, of the Cregganbaun, Pollanoughty, Oughty, Fiddaunarinnia, Lough Nacorra, and Owenwee Groups. The word Cregganbaun was also used by Stanton to describe all the Silurian rocks in the district, and it was used by Anderson for rocks that Dewey later named the Pollanoughty, Oughty, and Fiddaunarinnia Groups, inclusively. Dewey's Cregganbaun Group is apparently the same as Anderson's Croagh Patrick Group, and these geographic names were used by both authors in different senses. Finally, Dewey's Lough Nacorra and Owenwee Groups are apparently the equivalent of Anderson's Knockfadda Group.

The correlation of the Croagh Patrick "Series" is uncertain. Anderson (1960a, p. 269-270) collected some shelly fossils from his Cregganbaun Group, which were thought by A. Williams to be late Llandovery or Wenlock in age. Anderson believed that the beds were Wenlock because of their lithologic similarity to the Upper

Owenduff Group of the Galway area to the south. Dewey (1963, p. 337, 340) believed that his Croagh Patrick "Series" was a facies of the Upper Owenduff and Salrock Groups to the south. More recently, Phillips and others (1970, p. 207) correlated the Croagh Patrick "Series" with the Wenlock beds below the Salrock Formation of northwest Galway.

On nearby Clare Island, rocks similar to the Croagh Patrick "Series" occur and have been named the Knockmore Group (Hallissy, 1914).

CROCKNAGARGAN BEDS
(12)

See Little River Group.

CROSSRIDGE FORMATION
(22)

See Carmichael Burn Group.

CROUGHAN BEDS
(16)

Capewell (1957) distinguished sandstone and conglomerate east of the Comeragh Mountains from the underlying deep-water sediments. These beds are intensely folded, and Charlesworth (1963, p. 190) concluded that they were probably of late Ludlow or Downton age.

CWMERE GROUP
(36)

See Pont Erwyd "Stage."

CWM LANE BEDS
(48)

This is a thin unit that has not yielded fossils that could be diagnosed for correlation purposes. Potter and Price (1965, p. 395) regarded it as the probable equivalent of the lowest Red Marls to the northeast and therefore probably of Downton age.

CYRTOCERAS MUDSTONES
(53)

Straw (1937, p. 410) stated that "the Cyrtoceras Mudstones rest everywhere on graptolitic shales belonging to the zone of *Monograptus scanicus* and they contain a sparse graptolite fauna which appears to be referable to the same zone." However, Elles (1944) later stated that the forms reported from these units range throughout the *M. nilssoni* and *M. scanicus* zones.

DALMANELLA LUNATA BEDS OF SOUTHWEST CLUN FOREST AND KERRY
(56)

These beds have yielded a Whitcliffian Community (Earp, 1938, p. 134–135; 1940, p. 6) and are assigned to the Whitcliffian Stage because they overlie beds containing *Monograptus leintwardinensis* and underlie the *Platychisma helicites* Beds of Downton age.

DAMERY BEDS
(71)

These beds were described briefly by Curtis (1955b, p. 5-6) and contain the type locality of *Eocoelia curtisi* (Ziegler, 1966, p. 537). This fossil is particularly abundant in the lower part of the Damery Beds, and *Costistricklandia lirata* var. *alpha* occurs in the upper part (Ziegler and others, 1968b, p. 761). These occurrences indicate a C_5 correlation for the Damery Beds. The beds rest with slight angular unconformity on the Cambrian Micklewood Beds in the northern part of the Tortworth inlier, but elsewhere they rest on the Lower Trap; they are overlain by the Upper Trap, which is succeeded by the C_6 Tortworth Beds.

DARK GREY SLATY ROCK OF THE CLIFFE BOREHOLE
(78)

The original description of this borehole is contained in Whitaker (1908a, p. 383), who reported *Atrypa reticularis* and *Plectambonites* sp. Later, Bullard and others (1940, p. 89) pointed out the similarity of this material to the probable Wenlock-age rocks of the Ware borehole and relisted the fossils as *Atrypa reticularis* and *Leptelloidea (Leangella) segmentum*. Ziegler has recently re-examined the material preserved in the Geological Survey Museum in London and would list the fossils ?*Atrypa reticularis, Eoplectodonta* cf. *millinensis,* and *Clorinda* sp. These occurrences are consistent with a Wenlock or latest Llandovery age.

DARK SLATY ROCK OF THE SHEERNESS BOREHOLE
(77)

This hole was drilled with a percussion drill so that only fragments were obtained; these did not include fossils. Lamplugh and others (1923, p. 184) compared the fragments with the Silurian strata of the nearby Chilham borehole. The Sheerness borehole entered the ?Silurian to a depth of 11 meters below the unconformity with the Lower Greensand.

DAYIA NAVICULA BEDS OF SOUTHWEST CLUN AND KERRY
(56)

These beds are assigned, at least in part, to the zone of *Monograptus leintwardinensis* because of the occurrence of that species in the lower part of the formation in the southwest Clun area (Earp, 1940, p. 5).

DAYIA SHALES OF BUCKNELL
(55)

These beds were described by Stamp (1919, p. 226-227), who regarded them as resting on lower Ludlow Shale. Holland (1959b, p. 470-471), who described the adjacent Knighton area, disputed the lower Ludlow designation and considered both Stamp's Dayia Shales and lower Ludlow Shale to be the equivalent of the Lower to Middle Knucklas Castle Beds of Knighton. Holland further correlated the Dayia Shales of Bucknell with the lower part of the Dayia Navicula Beds of Kerry and southwest Clun. All these correlatives of the Bucknell Dayia Shales rest on beds containing *Monograptus leintwardinensis;* therefore, the presumption is that the Dayia Shales are of late Leintwardinian or Whitcliffian age.

DDOL SHALES
(40)

See Gwastaden Group.

DEERHOPE BURN FLAGSTONES
(23)

These flagstones have not yielded diagnostic fossils, but their stratigraphic position above the upper Llandovery Reservoir Beds and beneath beds E to H of early Wenlock age limits the age of the flagstones to the latest Llandovery or earliest Wenlock (Mitchell and Mykura, 1962, p. 15).

DENBIGH GRITS GROUP
(34)

P. T. Warren and his associates of the Institute of Geological Sciences of Great Britain have been revising the stratigraphy of northwest Denbighshire. Graptolites are rare in the Denbigh Grits Group, but Warren (1971, p. 452) believed that it ranges in age from the *Cyrtograptus centrifugus* to the *C. linnarssoni* zone. It is not clear whether the faunas he collected came from the equivalent Benarth Flags and Grits of Conway (col. 33) or from northwest Denbighshire. The Fynyddog Grits of Trannon (col. 38) are also part of this great lithological unit, as is the Pen-y-glog Grit that occurs within the Pen-y-glog Group at Llangollen (col. 35).

DERWEN GROUP
(36)

See Pont Erwyd "Stage."

DEVILS BRIDGE GROUP
(37)

See Ystwyth "Stage."

DINAS-BRAN GROUP
(35)

Wills and Smith (1922, p. 208) have reported some shelly fossils of lower Whitcliffe aspect, such as *Dayia navicula* and *Chonetes striatellus* (=*Protochonetes ludlovien-sis*), from this group. The Vivod Group, which contains *Monograptus leintwardinen-sis*, underlies the Dinas-bran Group, and the Carboniferous rests with pronounced angular unconformity on top.

DINGLE GROUP
(1)

The age of the Dingle Group has long been a matter of debate. It is unfossiliferous and was thought by Shackleton (1940, p. 10) to be separated from the underlying Croaghmarhin Formation of Ludlow age by an unconformity and from the overlying Old Red Sandstone by a pronounced angular unconformity. After correlating the orogenic episodes, represented by the angular unconformities above and beneath the beds, with unconformities elsewhere in Ireland and on the Continent, Shackleton (1940, p. 12) concluded that the Dingle Group might be of Downton (Pridoli) age. More recently, however, Holland (1969a, p. 306) has concluded that the lower unconformity does not exist; therefore, there is no precise indication of the age of the base of the Dingle Group. We follow Cocks and others (1971) in placing this contact arbitrarily within the upper Ludlow.

DIPPAL BURN FORMATION
(20)

See Waterhead Group.

DOLGADFAN GROUP
(38)

Wood (1906, p. 682–686) collected large graptolite faunas from this group and recognized two zones, the zone of *Monograptus fimbriatus*, which is equivalent

to the zone of *M. gregarius* of present usage, and the zone of *M. convolutus.*
The overlying Twymyn Group contains the zone of *Cephalograptus cometa* in
its lower beds, and this zone is usually taken as the upper part of the zone of
Monograptus convolutus; therefore, the contact between these two groups has
been placed within the zone of *M. convolutus.*

DOLGAU GROUP
(38)

See Tarannon "Series."

DOLYHIR LIMESTONE
(54)

See Nash Scar and Dolyhir Limestone.

DOUGLAS WATER ARENITE
(21)

See Glenbuck Group.

DOVESTONE REDBEDS
(21)

See Glenbuck Group.

DOWNTON CASTLE SANDSTONE GROUP OF BIRMINGHAM
(64)

Ball (1951, p. 226) has subdivided the Downton Castle Sandstone into the Tuner's
Hill Beds and the Gornal Sandstone, and he has included within the group the
thin but persistent Ludlow Bone Bed. These rocks are the lithological equivalent
of the lower half of the Grey Downton Formation of the nearby Ludlow area.

DOWNTON CASTLE SANDSTONE GROUP OF BUCKNELL
(55)

This area has not been studied in recent years and Stamp (1919, p. 231-234)
simply correlated these sandstone beds with the Downton Castle Sandstone of
the Ludlow area on lithological grounds (see Grey Downton Formation).

DOWNTON CASTLE SANDSTONE OF LUDLOW, CLEOBURY MORTIMER, AND THE MALVERN DISTRICT
(63, 65)

See Grey Downton Formation.

DROM POINT FORMATION
(1)

See Dunquin Group.

DRUMYORK FLAGS
(18)

Cocks and Toghill (1973) have collected the *M. griestoniensis* and *M. crenulata* zones from this unit.

DRYNGARN CONGLOMERATE
(42)

See "Lower Birkhill" of Abergwesyn and Pumpsaint.

DUDLEY LIMESTONE
(64)

Butler (1939, p. 55) was of the opinion that the Dudley Limestone is the exact correlative of the Wenlock Limestone of the type area, because he obtained *Monograptus flemingi* var. *delta* both from immediately above and immediately below the unit. This variety was reported by Das Gupta (1932, p. 351) from beneath the limestone of the type area in the *Cyrtograptus lundgreni* zone.

DUNGAVEL GROUP
(20)

The lower unit of this group, the Middlefield Conglomerate, "rests with paraconformity or slight angular unconformity" on the Logan Formation (Walton, 1965, p. 198) and is succeeded by the Plewlands Sandstone. Paleontological evidence for the age of this group is lacking, but the Middlefield Conglomerate contains an abundance of quartzite pebbles and can be correlated with similar conglomerates elsewhere in the Midland Valley (Rolfe, 1960, p. 254). In the Hagshaw Hills inlier, it is known as the Hareshaw Conglomerate (Rolfe, 1960, p. 242); in the Carmichael inlier it is called the Kirk Hill Conglomerate (Rolfe, 1960, p. 247), while in the North Esk inlier it occurs in the Lyne Water Fish Beds (Mitchell and Mykura, 1962, p. 12).

DUNQUIN GROUP
(1)

The Dunquin Group was first named by Holland (1969a, p. 301) and includes, from bottom to top, the Ferriters Cove Formation, the Clougher Head Formation, the Mill Cove Formation, the Drom Point Formation, and the Croaghmarhin Formation.

Gardiner and Reynolds (1902, p. 244) have reported *Pentamerus oblongus* and *Stricklandia lens* from the Ferriters Cove Formation, which would indicate a late Llandovery age, but Lamont (1965, p. 20) has doubted these identifications. The occurrence at many horizons of *Macropleura bijugosa* indicates that the beds are no older than C_3, and an early Wenlock age is provisionally accepted (Holland, 1969a, p. 302; Cocks and others, 1971).

The Clougher Head Formation is mostly volcanic, but some of the interbedded sediments contain a fauna similar to the underlying formation (Holland, 1969a, p. 304). The Mill Cove Formation is unfossiliferous. The Drom Point Formation has been redefined by Holland (p. 305) to include beds containing *Rhipidium* aff. *R. pingue* Amsden (1949) on Great Blasket Island; this suggests that the top of the Drom Point Formation is of late Wenlock-Ludlow age. The upper beds of the Croaghmarin Formation contain *Dayia navicula*, which suggests a Ludlow age for these horizons (Holland, 1969a, p. 306).

DUNSIDE FORMATION
(20)

See Priesthill Group.

DYFFRYN FLAGS
(40)

See Gwastaden Group.

EASTGATE FORMATION
(22)

This formation is barren (Rolfe, 1960, p. 252), and because both the underlying Kirk Hill Conglomerate and the overlying Old Red Sandstone are very poorly dated, it could be of Ludlow or Downton age.

EDEN VALE BEDS
(12)

See Little River Group.

EISTEDDFA GROUP
(37)

See Pont Erwyd "Stage."

EITHINEN BEDS
(31)

Woods and Crosfield (1925, p. 179) assigned these beds to the zone of *Monograptus tumescens*, an interval which, for the purposes of this chart, we place within the zone of *M. scanicus.* The beds do not yield an abundant graptolite fauna, but their date is assured by the fact that they occur above beds with a *M. scanicus* assemblage and below beds with a *M. tumescens* assemblage.

ELWY GROUP
(34)

The Elwy Group contains graptolite assemblages of the upper *Monograptus nilssoni*, the *M. scanicus*, and the *M. leintwardinensis incipiens* zones (Warren, 1971, p. 451).

FACHDRE GROUP
(38)

This is the earliest group seen in the Trannon area, and its base is unknown (Wood, 1906, p. 686-687). Wood reported *Dimorphograptus swanstoni, Climacograptus medius, C. hughesi, Diplograptus vesiculosus,* and *Monograptus tenuis* from the best locality, and applied the name "zone of *Dimorphograptus swanstoni*" to the assemblage. The presence of this species and *Diplograptus vesiculosus* indicates a correlation with the zone of *C. vesiculosus* of present usage. The overlying Dolgadfan Group contains *Monograptus fimbriatus* in its base, suggesting that this overlying group does not range below the *M. gregarius* zone.

FENCE CONGLOMERATE
(22)

Rolfe (1960, p. 254) has correlated this conglomerate with the Parishholm Conglomerate of the Hagshaw Hills and the Red Igneous Conglomerate of North

Esk on a lithologic basis. In all these areas, the conglomerate overlies beds of late Llandovery or early Wenlock age, and it underlies fish beds of late Wenlock or early to middle Ludlow age. The lower contact of these conglomerates has been interpreted as a paraconformity (Rolfe and Fritz, 1966, p. 163); therefore, these units may be of late Wenlock age.

FERRITERS COVE FORMATION
(1)

See Dunquin Group.

FIDDAUNARINNIA GROUP
(5)

See Croagh Patrick "Series."

FINE SANDSTONES OF THE CURLEW MOUNTAINS
(7)

Charlesworth (1960, p. 47) lists a brachiopod fauna identified by Williams, including *Stricklandia lens ultima*, which indicates a possible range of C_3 to C_5.

FISH BED FORMATION
(21)

See Glenbuck Group.

FOLLY SANDSTONE
(54)

Ziegler and others (1968b, p. 750) report *Eocoelia hemisphaerica* and *Stricklandia lens* aff. *progressa* from these sandstones, which would indicate a C_1-C_2 correlation. The base of the sequence is not visible, and the sandstones are overlain unconformably by the Nash Scar Limestone, which rests directly on Precambrian conglomerate and mudstone at Dolyhir (Kirk, 1951a, p. 56).

FRON FRYS SLATES
(35)

The Fron Frys Slates have not yielded diagnostic fossils from their type locality on the south side of the Llangollen syncline, but Wills and Smith (1922, p. 194-195) report abundant faunas representing the zones of *Monograptus cyphus, M. gregarius*, and *M. convolutus* from the presumed equivalents of these slates on the northern side of the syncline. The Fron Frys Slates rest on the Corwen Grit, which has been correlated with the Hirnant Limestone because of its similar lithology and shelly fauna (Wills and Smith, 1922, p. 191). These lines of evidence suggest that the Fron Frys Slates probably represent the whole of the lower and middle Llandovery intervals.

FYNYDDOG GRITS
(38)

These grits are the local representatives of the Denbigh Grits and have not yielded fossils (A. Wood, 1906, p. 653). They rest on the Nant-ysgollon Shales, which probably extend into the zone of *Monograptus riccartonensis*. The upper limit of Fynyddog Grits is unknown in this district.

GAFALLT BEDS
(41)

See Caban Group.

GALA (QUEENSBURY) GROUP
(25)

Unlike the underlying Birkhill Shales, the Gala (Queensbury) Group is sparsely fossiliferous. Moreover, because of the complicated structure and rapid facies changes (C. Lapworth, 1878b, p. 341), the stratigraphy of these rocks has never been properly understood. Peach and Horne (1899, p. 210) recognized three units within the upper Llandovery of the Moffat District: (1) Purple and gray flagstones and shales, (2) Queensberry Grits, and (3) Hawick Rocks, but the time ranges of these units probably overlap. Peach and Horne (1899, p. 212) listed *Monograptus exiguus, M. crispus, M. priodon, M. sedgwicki*, and *Rastrites maximus* from these rocks, all of which indicate a late Llandovery age.

"GALA" OF ABERGWESYN AND PUMPSAINT
(42)

At the base of the "Gala" of Abergwesyn and Pumpsaint, Davies (1933, p. 187) has collected a large graptolite fauna representing the zone of *Monograptus turriculatus. M. crispus* has not yet been found in the area, but Davies believed that the zone of *M. crispus* might be present in the rocks beneath the distinctive Pysgotwr Grits, because he consistently discovered a graptolite fauna of *M. dextrorsus* and *M. spiralis* above the horizons at which *M. turriculatus* occurred.

However, these species, including *M. turriculatus,* are known to range throughout the zones of *M. turriculatus* and *M. crispus.*

The top two lithologic units of the area are named the Pysgotwr Grits and the Green and Purple Shales. Davies assigned the Pysgotwr Grits to the zone of *M. griestoniensis* because of the occurrence of that species as well as *M. dextrorsus* and *M. proteus.* The Green and Purple Shales have yielded a *M. crenulatus* zone fauna. The top of the sequence in the area is unknown.

GARHEUGH FORMATION
(27)

Rust (1965a, p. 108) believed that this formation was equivalent to the zones of *Monograptus turriculatus* and *M. crispus* because of the occurrence of *M. exiguus.* This formation is separated by faults from the underlying Kilfillan Formation and the overlying Hawick Rocks.

GASWORKS MUDSTONE
(44)

See Haverford "Stage."

GASWORKS SANDSTONE
(44)

See Haverford "Stage."

GELLI GROUP
(38)

See Tarannon "Series."

GILGRIN MUDSTONES
(40)

See Gwastaden Group.

GLENBUCK GROUP
(21)

Of the four units that comprise the Glenbuck Group (Rolfe, 1962, p. 242), the Douglas Water Arenite, the Dovestone Redbeds, the Fish Bed Formation, and the Gully Redbeds, only the Fish Bed Formation has yielded datable fossils. These are of late Wenlock to middle Ludlow age, according to Westoll (1951, p. 7).

GLENCRAFF FORMATION
(4)

See Upper Owenduff Group.

GLEN DING FORMATION
(15)

See Shales, Grits, and Conglomerates of Slieve Bloom, Knockshigowna, Blessington, and Chair of Kildare.

GLENSHALLOCH SHALE
(17)

Cocks and Toghill (1973) have collected a large graptolite fauna from these beds that is indicative of the *R. magnus* subzone.

GLENWELLS CONGLOMERATE
(17)

This is a new stratigraphic unit proposed by Cocks and Toghill (1973), and these beds were originally included in the Newlands Sandstone by Lapworth (1882). They have not yielded a diagnostic fauna but are presumed to be basal middle Llandovery or top lower Llandovery from their stratigraphic position.

GLENWELLS SHALE
(17)

Cocks (1971, written commun.) wrote, "The top of the Woodland Fm., the Tralorg Fm. and the Glenwells Shale all carry the same graptolite fauna, indicative of the upper *M. cyphus* zone, including *M. cyphus* itself." The Girvan area has recently been revised by Cocks and Toghill (1973).

GLYN-DYFRDWY GROUP
(35)

Wood (1900, p. 446) has reported both the zones of *Monograptus vulgaris* (= *M. ludensis*) and *M. nilssoni* from these beds, which she termed the Nantglyn Flags.

The first of these zones was represented by the zone fossil itself and the second by its zone fossil and also *M. bohemicus*. Wills and Smith (1922, p. 204-205) agreed with these assignments.

GORNAL SANDSTONE
(64)

See Downton Castle Sandstone Group.

GORSLEY LIMESTONE
(67)

Lawson (1954, p. 233-236) has considered the problem of the correlation of the Gorsley Limestone. He pointed out that the lithologic and faunal data is more similar to the Wenlock Limestone than to the Aymestry Limestone, and that in any case, the Aymestry Limestone equivalents of May Hill to the southeast and Woolhope to the northwest pinch out when traced toward Gorsley. The correlation of the Gorsley Limestone with the Wenlock Limestone shows that the Ludlow succession is only 3.5 meters thick, but again the trend of thinning toward Gorsley can be observed in the Ludlow beds of May Hill and Woolhope.

GRAPTOLITIC SHALES OF THE SLIEVE BERNAGH INLIER
(3)

These beds were discovered by Rickards and Archer (1969) who collected graptolite faunas of the zones of *Akidograptus acuminatus*, *Monograptus atavus*, *Monograptus turriculatus*, and the subzone of *R. maximus*. The outcrop is apparently a fault slice, so that direct evidence of its stratigraphic relation with other units of the inlier is uncertain.

GRAPTOLITIC SLATES FROM THE CHILHAM BOREHOLE
(75)

This borehole penetrated the Silurian to a depth of 55 meters below the unconformity with the Lias, although the actual thickness represented may be only 6 meters because the beds dip at high angles (Lamplugh and others, 1923, p. 123, 130). These authors reported the following graptolites identified by Elles: *Monograptus crispus, M. discus, M. exiguus, M. marri, M. nodifer, M. nudus, M. priodon*, and *Rastrites equidistans*. This is, without doubt, a *Monograptus crispus* zone assemblage.

GREEN AND PURPLE SHALES
(42)

See "Gala" of Abergwesyn and Pumpsaint.

GREEN DOWNTONIAN BEDS
(54, 56, 59)

Earp (1938, p. 137-138) and Holland (1959b, p. 462-464) have described these beds, which contain the typical Welsh Borderland Downton faunal community. Holland believed that the Yellow Downtonian of Knighton was equivalent to the lower half of the Green Downtonian Beds of the adjacent Kerry and southwest Clun Forest area. This unit has also been reported from the Long Mountain area (Cocks and others, 1971, Fig. 2).

GREY DOWNTON FORMATION, INCLUDING LUDLOW BONE BED
(62, 63, 65, 66, 69)

The area around Ludlow is the type area for the Ludlow and Downton sequences; therefore, the Grey Downton Formation is the lower half of the Downton (=Pridoli) by definition. The subdivisions of the Grey Downton Formation are the Ludlow Bone Bed, the Downton Castle Sandstone, and the Temeside Beds (Dineley and Gossage, 1959, p. 222). The Ludlow Bone Bed is a very thin unit, but it is widely recognized because it occurs between two distinctively different formations. It is known from the Ludlow (Holland and others, 1963, p. 125), Wenlock (Ball and White, 1961, p. 180), Birmingham (Ball, 1951, p. 226), Malvern (Phipps and Reeve, 1967, p. 353), Usk (Walmsley, 1959, p. 485), Woolhope (Squirrell and Tucker, 1960, p. 154), and also at May Hill and Gorsley where it is called the Phosphatized Pebble-bed and Upper Phosphatized Pebble-bed, respectively (Lawson, 1954, p. 229; 1955, p. 86).

The Downton Castle Sandstone and the Temeside Beds are widely recognized and contain the fish zone *Hemicyclaspis* (White, 1950, p. 53; Allen and Tarlo, 1963, p. 131). These rocks have been described at Brown Clee Hill by Ball and White (1961, p. 180).

Martinsson (1967, p. 376) has identified the ostracodes *Frostiella groenvalliana* and *Londinia* cf. *kiesowi* from the lowermost Downton Castle Sandstone. These two species are the indices of one of Martinsson's ostracode zones.

GREY SANDSTONES OF THE CURLEW MOUNTAINS
(7)

These sandstones are unfossiliferous and have been correlated on a lithologic basis with the Upper Owenduff Group of northwest County Galway by Charlesworth (1960, p. 48).

GRITS OF DONAGHADEE
(9)

These grits succeed shales of middle Llandovery age and have been correlated on a lithologic basis with the Gala Series of Scotland (Swanston and Lapworth, 1878).

GRITS WITH BLACK SLATES AT BALBRIGGAN
(8)

Elles identified several graptolite assemblages for Gardiner (1899, p. 399) from various horizons in these beds. They were attributed to the *M. gregarius, M. spinigerus (sedgwicki),* and *M. riccartonensis* zones. Recent unpublished work by Rickards and Archer has established the following zones at Balbriggan, based upon numerous graptolite species: *A. acuminatus, M. triangulatus, D. magnus, M. sedgwicki, M. turriculatus, C. centrifugus, C. murchisoni, M. riccartonensis, C. linnarssoni,* and *C. lundgreni.* There is almost certainly a complete Llandovery and Wenlock suite of zones in a graptolite facies. The absence of some in the above list can be explained by the presence of small faults and some relatively barren horizons.

GULLY REDBEDS
(21)

See Glenbuck Group.

GUTTERFORD BURN FLAGSTONES
(23)

See Reservoir Beds.

GWASTADEN GROUP
(40)

This group, as presently conceived (Kelling and Woolands, 1969), consists of four units, the Cerig Gwynion Grits, the Dyffryn Flags, the Ddol Shales, and the Gilgrin Mudstones. The *A. acuminatus* zone occurs in the base of the Dyffryn Flags (H. Lapworth, 1900, p. 76), and the *M. cyphus* and *M. fimbriatus* (=*gregarius* in part) zones occur in the Ddol Shales. The *M. convolutus* zone has been reported from the Gilgrin Mudstones. *M. leptotheca* and *D. magnus* (H. Lapworth, 1900, p. 8) have also been reported, probably indicating the presence of those zones.

GWERN-Y-BRAIN GROUP
(58, 59)

This group apparently spans the Ashgill-Llandovery boundary, because Wade (1911, p. 430) reported a *Glyptograptus persculptus* zone fauna, including the name species, *Diplograptus* cf. *parvulus, Orthograptus truncatus* var. *socialis,* and such typical Ashgill forms as *Trinucleus* cf. *seticornis.*

GYFFIN SHALES
(33)

Elles (1909, p. 184–186) reported the following zones represented by abundant graptolite faunas: *Monograptus gregarius, M. sedgwicki, Rastrites maximus,* and *M. crispus.* In addition, she reported a fauna representing a distinctly earlier horizon than any of the above zones; it included *Mesograptus modestus, Climacograptus normalis, C. rectangularis, C. hughesi,* and *Glyptograptus persculptus.* However, it is difficult to prove that the Gyffin Shales include the lowest zone of the Llandovery, and in fact, Elles felt that the underlying unit, the Conway Castle Grits, formed the base of the Llandovery succession. She correlated the Conway Castle Grits with the Hirnant Limestone (Elles, 1909, p. 183) on the basis of a similar shelly fauna, and the Hirnant Beds were later shown by Pugh (1929, p. 265) to underlie the lowest Llandovery graptolite zone, the zone of *Glyptograptus persculptus.* As for the top of the Gyffin Shales, Elles (1909, p. 186) stated that they have yielded the characteristic graptolites belonging to the zone of *Monograptus crenulatus,* but she did not list these.

HAGGIS GRIT
(23)

This unit is poorly fossiliferous, but may be dated from the fact that it underlies beds E to H of early Wenlock age and overlies both Deerhope Burn Flagstones and Reservoir Beds, the latter containing C_6 fossils (Mitchell and Mykura, 1962, p. 16).

HAGSHAW GROUP
(21)

The Hagshaw Group includes a lower unit, the Smithy Burn Siltstone, and an upper unit, the Ree Burn Formation (Rolfe, 1962, p. 242). Brachiopods from the Ree Burn Formation are suggestive of a late Llandovery or early Wenlock age (Rolfe, 1962); *Atrypa reticularis* and *Howellella* sp. establish a lower limit of C_2 or C_3, and *Pentlandella haswelli* is not known above beds of early Wenlock age. The base of the Smithy Burn Siltstone is unknown, and the only fossils of correlative value discovered in these beds were identified by an undergraduate, J. D. Wilson, and listed in his thesis at the University of Glasgow. Unfortunately, the specimens were discarded before being identified professionally, but I. Strachan has interpreted the list (Rolfe and Fritz, 1966, p. 160), which contains *Monograptus vomerinus* cf. var. *crenulatus* and *M. griestoniensis, M. spiralis, M. priodon,* and *M. marri,* as indicating a post-*M. crispus* zone of late Llandovery age.

HARESHAW CONGLOMERATE
(21)

This unit is similar lithologically to the Middlefield Conglomerate of Lesmahagow, to the Kirk Hill Conglomerate of Carmichael, and to a conglomerate in the Lyne Water Fish Beds of North Esk (Rolfe, 1960, p. 254; 1962, p. 242). Fish beds below this conglomerate in the Hagshaw Hills have been dated as having a possible range in age of late Wenlock to middle Ludlow (Westoll, 1951, p. 7).

HAUGH WOOD BEDS
(66)

Squirrell and Tucker (1960, p. 142) defined the Lower Haugh Wood Beds, whose base is unknown, and the Upper Haugh Wood Beds, a relatively thin series just beneath the Woolhope Limestone. They report *Palaeocyclus* sp. from the Upper Haugh Wood Beds and *Stricklandia lirata* from both Upper and Lower Haugh Wood Beds. Ziegler and others (1968b, p. 757) have determined the variety of stricklandiid from the Lower Haugh Wood Beds to be *Costistricklandia lirata* var. *typica*. These occurrences indicate a C_6 age for both Lower and Upper Haugh Wood Beds, and the single graptolite occurrence in the Lower Haugh Wood Beds of *Retiolites geinitzianus* var. *angustidens* is in accord with this (Squirrell and Tucker, 1960). The *Petalocrinus* Limestone is present in the highest 3 meters of the Upper Haugh Wood Beds; it was found by Pocock (1930, p. 60–61) at equivalent horizons in the May Hill and Malvern areas.

HAVERFORD "STAGE"
(44)

Four units comprise the Haverford "Stage": the Basement Beds, the Cartlett Beds, the Gasworks Mudstone, and the Gasworks Sandstone (Strahan and others, 1914, p. 81). The Basement Beds have not yielded diagnostic fossils, and they rest on the Slade Beds of Ashgill age. The basal Cartlett Beds have yielded *Tretaspis* sp., which Cocks (1968, p. 303–304) interprets as indicative of an age close to the Ashgill-Llandovery boundary. A single graptolite specimen, identified as ?*Diplograptus* (*Mesograptus*) *modestus* var. *parvulus*, has been discovered from a low horizon in the Cartlett Beds (Cocks, 1968, p. 89). This variety is possibly restricted to the zone of *Akidograptus acuminatus* at the base of the Llandovery. Also, it should be pointed out that Williams (1951, p. 128–129) regarded the basal Cartlett Beds, sometimes known as the St. Martin's Cemetery beds, as equivalent to the unfossiliferous A_1 sandstones of the Llandovery area, and he actually incorporated the list of species from the St. Martin's Cemetery beds in his table, which shows the distribution of brachiopods in the Llandovery. Boucot has collected *Stricklandia lens* cf. *typica* from the Gasworks Mudstone (U.S. Natl. Museum loc. no. 10511), which would suggest a correlation with A_3-A_4 for that unit. Finally, the upper beds of the Gasworks Sandstone have yielded a single specimen of *Climacograptus scalaris* var. *normalis*, which indicates an upper limit of the zone of *Monograptus gregarius*.

HAWICK ROCKS
(26, 27)

Rust (1965a, p. 106) subdivided the Hawick Rocks of the Whithorn area into a lower unit, the Carghidown Beds, and an upper unit, the Kirkmaiden Beds, on the basis of the presence or absence of redbeds, respectively. Rust (1965a, p. 105) discovered graptolites in the Kirkmaiden Beds, which Strachan described as "slender monograptids of the vomerinus type" and said "the fauna gives the general impression of the vomerinids occurring at the top of the Valentian (*grieston-iensis* and *crenulatus* zones)." The Hawick Rocks of the Whithorn area are separated from the underlying and overlying formations by faults.

The Hawick Rocks of the Hawick area have not yielded diagnostic fossils and were thought by Warren (1964, p. 208) to be facies of the Riccarton Group. The structure of this area is rather difficult, however, and we follow Rust in provisionally correlating these rocks with the upper Llandovery. One of us (Rickards) has examined the specimens and believes that they could equally well be early Wenlock in age.

See also the notes under "Gala Group."

HOLDGATE SANDSTONES
(63)

See Red Downton Formation.

HOLOPELLA GRITS AND SHALES
(53)

Diagnostic fossils have not been obtained from these beds, but they overlie the Orthonota Mudstones, which contain *Monograptus leintwardinensis* and underlie the Downton (Straw, 1937).

HORIZON V_1 OF MEIFOD
(57)

A few shelly fossils have been discovered in this unit (King, 1928, p. 685–687) but are not diagnostic as to age. However, the underlying Ashgillian unit has yielded graptolites normally associated with the *Dicellograptus anceps* zone, and the overlying horizon V_2 has yielded *Climacograptus normalis* and ?*Monograptus acinaces*, species that do not range above the lower Llandovery. Horizon V_1 is therefore probably of early Llandovery age. Williams (1951, p. 129) stated that the V_1 faunas are undoubtedly comtemporaneous with the A_1 beds of the Llandovery district, and he included the elements of the V_1 fauna in his A_1 list. The V_1 beds rest disconformably on the Ashgillian (King, 1928, p. 682).

HORIZON V_2 OF MEIFOD
(57)

In the lower part of this unit, *Climacograptus normalis* "and what Dr. G. L. Elles thinks may be a specimen of *Monograptus acinaces*" have been found (King, 1928, p. 688). These species would suggest the zones of *Monograptus acinaces* or *cyphus*, but the unit probably extends into the late Llandovery because the lowest graptolite fauna in the overlying horizon V_3 is in the zone of *M. turriculatus*. The higher beds of horizon V_2 contain an abundant shelly fauna, including *Meifodia subunta* forma *typica*, and on this basis, Williams (1951, p. 109) correlated these beds with the lower Llandovery A_{3-4} beds of the type area. The fauna of this locality has recently been redescribed by Temple (1970, p. 4), who agrees with Williams' age assignment.

HORIZON V_3 OF MEIFOD
(57)

King (1928, p. 700) collected graptolite faunas from successive levels in this unit and recognized the zones of *Monograptus crispus*, *M. griestoniensis*, and *M. crenulatus*. His *M. crispus* zone fauna included the index species and *M. discus*, *M. marri*, and *M. planus;* his *M. griestoniensis* zone fauna included the index species and *M. marri, M.* aff. *priodon*, and *M. spiralis;* his *M. crenulatus* zone fauna included the index species, *M.* aff. *priodon*, and *Retiolites geinitzianus* var. *angustidens*. Several localities are represented in each of these three lists. In addition, there are beds in this unit that are below the *M. crispus* zone, and King surmised that these were the equivalent of the *M. turriculatus* zone; in a footnote to the paper he stated, "Fossils indicative of this zone have now been found."

HORTON FLAGS
(30)

The Horton Flags have yielded *Monograptus colonus, M. roemeri*, and *M. bohemicus* (King and Wilcockson, 1934, p. 23); they suggest the zone of *M. nilssoni*.

HUGHLEY (OR PURPLE) SHALES
(62)

These beds were described by Whittard (1928, p. 747-752) as the Purple Shales, but they were later renamed by Pocock and others (1938, p. 105-106, 109-110).

Ziegler and others (1968b, p. 749) have reported *Eocoelia intermedia, E. curtisi, E. sulcata*, and *Costistricklandia lirata* from different localities. These occurrences indicate a range of C_4 to C_6 and are in accord with the graptolite evidence recently reviewed by Cocks and Rickards (1969; see also Cocks and Walton, 1968, p. 395). The Hughley Shales rest on the Pentamerus Beds but overlap onto the Ordovician at the Onny River section.

HUNTLEY HILL BEDS
(68)

These beds have yielded *Eocoelia hemisphaerica* from their middle horizons and *E. intermedia* from their higher horizons (Ziegler and others, 1968b, p. 759).

They therefore range from C_1 to C_4 and may include some middle Llandovery horizons. The relation of the Huntley Hill Beds to older rocks is difficult to determine; at Huntley Quarry, volcanic rocks occur and their dip would suggest an angular unconformity if no fault intervenes. These volcanic rocks could be the Precambrian Warren House Group of the Malverns. The contact of the Huntley Hill Beds with the overlying Yartleton Beds is obscured; it could be transitional or an unconformity.

INTERMEDIATE SHALES
(41)

See Caban Group.

KENLEY GRIT
(62)

Whittard (1928, p. 738-741) referred to this unit as the arenaceous beds, but Pocock and others (1938, p. 105-107) preferred the use of local names rather than lithologic names. The Kenley Grit rests unconformably on various Ordovician and Cambrian horizons. Diagnostic fossils have not been found in this unit, but it is directly overlain by the Pentamerus Beds of late middle Llandovery age.

KILBRIDE FORMATION
(4)

See Lower Owenduff Group.

KILFILLAN FORMATION
(27)

Rust (1965a, p. 108) stated that the Kilfillan Formation was the equivalent of the zones of *Akidograptus acuminatus*, *Orthograptus vesiculosus*, and *Monograptus gregarius* but did not give the evidence for this assignment. This formation is in fault contact with the later Garheugh Formation.

KILLARY HARBOUR GROUP
(4)

This group contains the Lough Muck Formation and the Salrock Formation. The Lough Muck Formation contains *Eocoelia angelini* and *Monoclimacis flumendosae* (Laird and McKerrow, 1970, p. 299), and these fossils are consistent with a middle Wenlock age. Presumably the Salrock Formation is late Wenlock or Ludlow.

KIP BURN FORMATION
(20)

See Priesthill Group.

KIRKBY MOOR FLAGS
(29)

The Kirkby Moor Flags contain shelly fossils and ostracodes of the Whitcliffe Beds of Ludlow (Shaw, 1971a, p. 371; Shaw, 1971b, p. 607-609).

KIRK HILL CONGLOMERATE
(22)

This conglomerate is unfossiliferous and has been correlated on a lithologic basis (Rolfe, 1960, p. 254) with the Middlefield Conglomerate of Lesmahagow, the Hareshaw Conglomerate of the Hagshaw Hills, and with a conglomerate in the Lyne Water Fish Beds of North Esk. Fish beds below this conglomerate in the Hagshaw Hills have been dated as late Wenlock or early to middle Ludlow (Westoll, 1951, p. 7).

KIRKMAIDEN BEDS
(27)

See Hawick Rocks.

KNOCKGARDNER FORMATION
(18)

This unit has not yielded diagnostic fossils but is presumed to be basal Wenlock because of its position directly above uppermost Llandovery beds.

KNOCKNAVEEN GROUP (INCLUDING KNOCKNAVEEN SANDSTONE FORMATION, KNOCKNAVEEN SILTSTONE FORMATION, AND KNOCKNAVEEN PEBBLY ARKOSE FORMATION)
(6)

This group has recently been defined to include, in ascending order, the above three formations (Phillips and others, 1970). The group rests conformably on the Toormore Group, and the top is not seen. Phillips and others (1970, p. 208) suggested that the Knocknaveen Pebbly Arkose Formation may be the lateral equivalent of the post-middle Wenlock Salrock Formation of northwest Galway. The only fossil reported from the Knocknaveen Group is "a poorly preserved *Lingula.*"

KNUCKLAS CASTLE BEDS
(54)

These beds are poorly fossiliferous (Holland, 1959b, p. 453) and overlie the Bailey Hill Beds, which have yielded *Monograptus leintwardinensis* at their summit. Holland and others (1963, Table II) have correlated the Knucklas Castle Beds with the Upper Leintwardine Beds and the Lower Whitcliffe Beds of the type area.

LADY BURN CONGLOMERATE
(17)

This name has been proposed by Cocks and Toghill (1973) and refers to the basal conglomerate of the Mulloch Hill Formation. The Lady Burn Conglomerate rests unconformably on the Ashgillian Drummock Group and High Mains Sandstone (Cocks and Toghill, 1973). The Drummock Group has yielded a *Dicellograptus anceps* zone fauna (Lamont, 1935, p. 301).

LAUCHLAN FORMATION
(18)

These beds have yielded a *M. griestoniensis* zone assemblage (Cocks and Toghill, 1973).

LEAZE FORMATION
(20)

See Waterhead Group.

LEDBURY GROUP
(65)

See Red Downton Formation.

LETTERGESH FORMATION
(4)

See Upper Owenduff Group.

LIME HILL BEDS
(12)

See Little River Group.

LIMESTONE OF THE CURLEW MOUNTAINS
(7)

This limestone contains a few corals that are not diagnostic for correlation purposes (Charlesworth, 1960, p. 47). The fine sandstones directly above the limestone have yielded an upper Llandovery brachiopod fauna; therefore, it is concluded that the limestone is of late Llandovery age.

LISBELLAW CONGLOMERATE
(11)

This conglomerate, which is not continuous along strike, contains occasional thin wedges of mudstone that have yielded graptolites (Harper and Hartley, 1938, p. 75). These represent the *Monograptus gregarius* zone and include *M. gregarius*, *M. triangulatus*, *Orthograptus bellulus*, and *Climacograptus* sp.

LITTLE RIVER GROUP
(12)

Fearnsides and others (1907, p. 110-111) have subdivided this group into five stratigraphic units, each characterized by an assemblage of graptolites. They correlated these local assemblage zones with the standard zones elsewhere in the British Isles. At the base of the group, the Crocknagargan Beds are characterized by the "Zone of *Cephalograptus acuminatus*," the basal zone of the Llandovery. Above this occur the Slate Quarry Beds with the "Zone of *Diplograptus modestus*," which they correlated with the zone of *Orthograptus vesiculosus* of Moffat (Fearnsides and others, 1907, p. 126). The Edenvale Beds follow and contain the "Zone of *Monograptus tenuis*," which was matched with the zone of *M. gregarius*. At that time the zone of *M. gregarius* was understood to include faunas up to the base of the zone of *M. sedgwicki*. The fourth unit described is the Mullaghnabuoyah Beds characterized ty the "Zone of *M. triangulatus*," which was also correlated with the zone of *M. gregarius*. Finally, the Lime Hill Beds contain the "Zone of *M. sedgwickii* [sic]," a widely recognized zone at the base of the upper Llandovery. Members of the Institute of Geological Sciences of Great Britain are revising the stratigraphic record of the region.

LLANIDLOES "STAGE"
(39)

Four units constitute this "stage": the Oldchapel Mudstones, the Caerau Group, the Moelfre Group, and the Pale Shales (W.D.V. Jones, 1945, p. 321-324). In the Oldchapel Mudstones, Jones recognized two zones, *Monograptus sedgwicki* and *M. halli* (both subzones of the *M. sedgwicki* zone of present usage) and also the band of *Rastrites maximus*. *M. crispus* has been discovered in the overlying Caerau Group, indicating the presence of the *M. crispus* zone (p. 322), and a small fauna containing *M. griestoniensis* was discovered in the Moelfre Group, indicating the zone of that species (p. 326). At the top of the Llanidloes "Stage," the Pale Shales have yielded *M. vomerinus* var. *crenulatus*, *M. marri*, *M. priodon*, and *M.* cf. *flagellaris*, which Jones (p. 327) referred to the zone of *M. crenulatus*.

LLAN-WEN HILL BEDS
(54)

These beds occur just beneath *Platyschisma helicites* beds that have been correlated with the Downton (Holland and others, 1963, Table II). They have yielded a typical upper Ludlow shelly community (Holland, 1959b, p. 454).

LLETTY BED FACIES
(48)

See Black Cock Beds.

LOGAN FORMATION
(20)

See Waterhead Group.

LONG MOUNTAIN SILTSTONE FORMATION
(59)

This formation was named in a brief account (Palmer, 1970a) and contains the *Monograptus nilssoni*, *M. scanicus*, *M. leintwardinensis incipiens*, and *M. leintwardinensis* zonal assemblages. It contains the Pentre Member, although the exact horizon of this member apparently is not clear.

LONG QUARRY BEDS
(48, 49, 50, 51, 52)

Potter and Price (1965, p. 388, 391) described this unit and followed Straw (1929) in assigning it to the base of the Downtonian because of its unconformable relations to the Ludlow rocks. As Potter and Price pointed out, authors in the past have varied considerably in placing the Ludlow-Downton contact in this succession because the faunas and lithologies differ to a certain degree from those of the type areas. Virtually all the fossils collected from so-called Ludlow and Downton rocks are shallow marine benthic types, so there is little hope at the present time of correlating the beds with any certainty. Richardson and Lister (1969, p. 211) identified spores from these beds that are consistent with a Ludlow or early Downton age.

LOUGH ACANON PELITIC FORMATION
(13)

This formation was described by Phillips and Skevington (1968) who collected Caradocian and lower Llandoverian (zone of *Akidograptus acuminatus*) graptolites from it.

LOUGH MASK FORMATION
(4)

See Lower Owenduff Group.

LOUGH MUCK FORMATION
(4)

See Killary Harbour Group.

LOUGH NACORRA GROUP
(5)

See Croagh Patrick "Series."

"LOWER BIRKHILL" OF ABERGWESYN AND PUMPSAINT
(42)

Davies (1933, p. 178-180) recognized five zones in his "lower Birkhill" which are, from the base, the zones of *Glyptograptus persculptus, Akidograptus acuminatus, Monograptus atavus, M. acinaces,* and *M. cyphus.* This is the same zonal sequence recognized elsewhere in central Wales in lower Llandovery rocks. The prominent Dryngarn conglomerate (Davies, 1926, p. 442, Fig. 2) occurs within these beds. It lies a little way below the top of the *M. atavus* zone and pinches out to the north and south.

LOWER BIRKHILL SHALES
(25)

See Birkhill Shales.

LOWER BLAISDON BEDS
(68)

Lawson (1955, p. 104) compared both Lower and Upper Blaisdon Beds with the Dayia Beds (Lower Leintwardine Beds) of Shropshire on general lithologic and faunal grounds. The Lower Blaisdon Beds have a basal limestone conglomerate and rest unconformably on the Upper Flaxley Beds.

LOWER BODENHAM BEDS
(66)

Squirrell and Tucker (1960, p. 159, 173) reported *Monograptus leintwardinensis* from these beds and compared them to the Dayia Shales (Lower Leintwardine Beds) of Ludlow, the Upper Blaisdon Beds of May Hill, and the Upper Llanbadoc Beds of Usk, because of the abundance of *Dayia navicula, Isorthis orbicularis,* and *Sphaerirhynchia wilsoni.* The basal Lower Bodenham Beds contain conglomeratic bands and in places rest directly on the Lower Sleaves Oak Beds. Fahraeus (1969, p.27) has interpreted the conodonts reported from these beds (Squirrell and Tucker, 1960, p. 178) as a *siluricus* zonal assemblage.

LOWER BRINGEWOOD BEDS
(62, 63)

This unit was named by Holland and others (1959) and was probably included by earlier workers in the Aymestry Limestone. *Monograptus tumescens* has been recorded from the Ludlow area, and in the Wenlock area these beds have yielded *M. chimaera chimaera, M.* cf. *salweyi, M. leintwardinensis incipiens, M. colonus colonus,* and *M. varians* (Shergold and Shirley, 1968, p. 125). We therefore correlate this unit with the *M. scanicus* zone.

LOWER BROWGILL BEDS
(29)

See Stockdale Shales.

LOWER CABAN CONGLOMERATE
(41)

See Caban Group.

LOWER CAMREGAN GRITS
(17, 18)

Cocks and Toghill (1973) have collected *Eocoelia curtisi* from these beds, indicating a C_4 or C_5 age.

LOWER COLDWELL FLAGS
(29)

See Upper Coniston Flags.

LOWER CONISTON FLAGS
(29)

The lower Coniston, or Brathay Flags, of the Howgill Fells range in age from the zone of *Cyrtograptus murchisoni* to the zone of *C. lundgreni,* according to Watney and Welch (1910, p. 473). These zonal identifications are supported by long faunal lists in a later publication (Watney and Welch, 1911, p. 217-227), and two intervening zones were also recognized, the zones of *Monograptus riccartonensis* and *Cyrtograptus rigidus;* this latter zone was thought to be equivalent to three zones of the Welsh Borderland, the zones of *C. symmetricus, C. linnarssoni,* and *C. rigidus* (Watney and Welch, 1911, p. 234). In any case, all of the Wenlock zones are represented in the Brathay Flags, up to and including the zone of *C. lundgreni,* although this also occurs in the overlying and partly equivalent Lower Coldwell Beds (Rickards, 1969). Recent work by Rickards (1964a, 1967, 1969) has refined the zonal schemes of Watney and Welch (1911) and Blackie (1933) in the Howgill Fells and Lake District, respectively, and has added considerably to the number of graptolite species and zones recorded.

The Brathay Flags have also been recognized in the Cross Fell inlier (Shotton, 1935, p. 661). Burgess and others (1970) have reported a *Cyrtograptus centrifugus* zone fauna from a locality thought by Shotton to be in the *M. crenulata* zone. On Keisley Beck they discovered a *C. linnarssoni* zone fauna.

LOWER CONISTON GRITS
(29)

See Coniston Grits.

LOWER CWM CLŶD BEDS
(50, 51, 52)

The Lower Cwm Clŷd Beds contain what Potter and Price (1965, p. 399) termed a "modified Upper Bringewood Bed fauna," that is, *Atrypa reticularis, Camarotoechia nucula, Dayia navicula, Isorthis orbicularis, Leptaena rhomboidalis,* and *Sphaerirhynchia wilsoni.* Potter and Price (p. 385) commented on the similarity of the Cwm Clŷd Beds of the northern Llandovery area to the Wilsonia Shales of Builth.

LOWER ELTON BEDS
(62, 63)

Holland and others (1959) gave the present name to this unit, which was formerly called the "Barren Beds" of the Lower Ludlow Shale; it has not yet yielded a diagnostic graptolite assemblage but has yielded probable *M. varians,* a *P. nilssoni* zone species. The Lower Elton Beds rest on the Wenlock Limestone, which has recently been reaffirmed as the uppermost unit of the Wenlock (Warren and others,

1966, p. 466). They underlie beds containing the *P. nilssoni* and *M. scanicus* zones and may themselves have a *P. nilssoni* zone fauna (Holland and others, 1969).

LOWER FLAXLEY BEDS
(68)

These beds rest on the Wenlock Limestone and are similar to the Lower Elton Beds of the Ludlow area (Holland and others, 1963, p. 149) that occur in the same stratigraphic position. Lawson (1955, p. 103) reported ?*Monograptus uncinatus* var. *orbatus* from the Lower Flaxley Beds, which suggests that they may in part be equivalent to the Middle Elton Beds, or, indeed, to the Lower Elton Beds of probable *P. nilssoni* zone age (see Lower Elton Beds).

LOWER FOREST BEDS
(69)

These beds rest on the Wenlock Limestone and contain, in their upper part, the graptolite *Monograptus* cf. *tumescens* (Walmsley, 1959, p. 507-508), indicating a correlation with the Eltonian or Lower Bringewoodian of the Ludlow area. Walmsley compared the Lower Forest Beds with the Lower Flaxley Beds of May Hill.

LOWER LEINTWARDINE BEDS OF TORTWORTH AND NEWNHAM
(71)

Mudstone and fine-grained sandstone in the Brookend borehole at Tites Point and at Newnham have been correlated with the Lower Leintwardine Beds of Ludlow (Cave and White, 1971, p. 243-249; Curtis, 1972, p. 29).

LOWER LEINTWARDINE BEDS OF WENLOCK AND LUDLOW
(62, 63)

This unit yields the zonal graptolite *M. leintwardinensis* (Holland and others, 1959) as do the Upper Leintwardine Beds. The Lower Leintwardine Beds were included in the Dayia or Mocktree Shales by earlier workers. In the Wenlock area, *M. leintwardinensis* occurs in association with *M. leintwardinensis incipiens* (Shergold and Shirley, 1968, p. 127).

LOWER LLANBADOC BEDS
(69)

These beds are the local equivalents of the Upper Bringewood Beds (Aymestry Limestone) of the Ludlow area (Walmsley, 1959, p. 508-509). Walmsley reported *Monograptus* cf. *leintwardinensis* var. *incipiens* from the Lower Llanbadoc Beds.

LOWER LLANDOVERY OF LLANDOVERY AND GARTH
(50, 51, 52, 53)

Considerable variation exists in the stratigraphy in the type area; therefore, it seems desirable to pick a section within the area to be used as a standard. The central part of the southern area of the Llandovery district, that is, the area about the River Ydw, is the most complete, most detailed, and the first one described (Jones, 1925b). In the southern part of the Llandovery district, Jones mapped four units in the lower Llandovery, which he named A_1, A_2, A_3, and A_4. Unfortunately, none of these units can be tied directly to the graptolitic sequence.

Andrew's (1925) description of the Garth area appeared at the same time as Jones's first Llandovery paper and has implications on the correlation of the Llandovery units with the graptolitic sequence. Andrew recognized three units in the lower Llandovery, A_a, A_b, and A_c. He correlated these with the first three of Jones's units, A_1, A_2, and A_3, respectively, on lithologic criteria (Andrew, 1925, p. 403). Further, he collected a graptolite fauna from a point 6 meters beneath A_a that consisted of *Glyptograptus* cf. *persculptus*, and *Diplograptus* (*Mesograptus*) cf. *modestus* var. *parvulus*. Therefore, the zone of *Glyptograptus persculptus* is indicated. Andrew realized this but preferred to put the boundary of the Bala and the Llandovery at the distinct lithologic change. Also, Andrew collected *Climacograptus* cf. *scalaris* and *Monograptus atavus* from his unit A_c.

Some information on the correlation of the top of the lower Llandovery sequence with the graptolitic sequence may be gained from the northern part of the Llandovery district, which was described by Jones in 1949 and also referred to in his 1925 paper. Jones mapped three units in this area, A_a, A_b, and A_c; these terms were meant to be noncommittal and did not imply a correlation with the contiguous southern area or with the Garth area. Jones (1925b, p. 360; 1949, p. 49) described a graptolite fauna from the highest beds of A_c that consisted of *Monograptus incommodus*, *Climacograptus hughesi*, *C. medius*, and *C. normalis?*. Jones correlated this fauna with the "uppermost part of the zone of *Monograptus acinaces*, near the junction with the overlying *M. cyphus* zone." Jones (1925b, p. 360) clearly believed that the uppermost A_c beds of the northern area were the equivalent of the A_4 beds of the southern area.

One other graptolite fauna should be mentioned, which is from the northeastern part of the southern Llandovery area. Jones (1925b, p. 357) collected *Climacograptus tornquisti* from beds that he assigned, questionably, to his unit A_2. This would imply a range of graptolite zones *M. atavus* to *M. sedgwicki* for the beds in question, but unfortunately these beds could as easily be A_3 or A_4 equivalents, so the graptolite information is not too useful.

Jones (1925b, p. 356) mapped his A_2 unit in the Sefin River section as resting on Bala rocks and overlain unconformably by C_1 of the upper Llandovery. These beds, however, are lithologically dissimilar to the typical A_2 mudstone because much sandstone is present and the correlation is uncertain.

Llandovery is the type area for the subspecies of *Stricklandia lens*. Williams (1951, p. 99–100) reported *S. lens prima* from A_b of the northern area and *S. lens typica* from A_3 and A_4 of the southern area.

LOWER LLANGIBBY BEDS
(69)

Holland and others (1963, p. 148) stated that the upper portions of the Ludlow sequence of the type area and Usk are similar in fauna, lithology, and thicknesses, and they match the Lower Llangibby Beds with the Upper Leintwardine Beds of the type area. Walmsley (1959, p. 512) reported *Monograptus* cf. *chimaera* from these beds. Also, Walmsley pointed out the similarity of the Lower Llangibby Beds with the Lower Longhope Beds of May Hill. The Institute of Geological Sciences of Great Britain Memoir of the Newport district (Squirrel and Downing, 1969, p. 31) listed *M. leintwardinensis* from these beds.

LOWER LONGHOPE BEDS
(68)

Holland and others (1963, p. 149) correlated this unit with the Upper Leintwardine Beds of Ludlow. In support of this correlation, they announced the discovery of *Chonetoidea grayi* in the Lower Longhope Beds; also, Lawson (1955, p. 105) had reported one specimen of *Neobeyrichia lauensis*. The Lower Longhope Beds pinch out to the north at Clifford's Mesne (Lawson, 1955, p. 106).

LOWER LUDLOW FORMATION OF THE MALVERN DISTRICT
(65)

This unit has recently been redescribed (Phipps and Reeve, 1967, p. 346–347) and may be tentatively correlated with the Eltonian Stage of the Ludlow district on general lithologic and faunal grounds.

"LOWER LUDLOW" GRAPTOLITIC SHALES OF BUILTH AND CLUN FOREST
(53, 54, 56)

Wood (1900, p. 431-438) recognized two graptolite zones in these shales, the zone of *Monograptus vulgaris* (= *Monograptus ludensis*) and the zone of *M. nilssoni*. She applied the letter B to beds containing graptolites of the lower zone, and C to beds containing graptolites of the higher zone. Only *M. vulgaris* and *M. dubius* were reported from the first zone, and *M. colonus*, *M. bohemicus*, *M. nilssoni*, and *Retiolites spinosus* were held to be characteristic of the *M. nilssoni* zone. The *M. ludensis* zone is now regarded as the top Wenlock zone; this formation apparently spans the Wenlock-Ludlow boundary.

Holland (1959b, p. 451) reported *M. colonus* and *M. nilssoni* from the Lower Ludlow Graptolitic Shales of Knighton, which indicate the presence of the *M. nilssoni* zone. These species, plus *M. bohemicus*, were found in this unit in the southwest part of the Clun Forest (Earp, 1940, p. 2), and a rich *M. nilssoni* fauna has been reported in the Kerry district (Earp, 1938, p.129).

At Presteigne, Kirk (1951b, p. 72) termed these beds the "Olive Mudstones" and stated that they may represent part of the zones of *C. lundgreni* and *M. nilssoni* as well as the *M. vulgaris* mudstones at Builth, which they greatly exceed in thickness. No faunal lists are given.

It should be mentioned that, in all these areas, the overlying units also apparently contain a *M. nilssoni* assemblage.

"LOWER LUDLOW" OF CARDIFF
(70)

See Silurian of Cardiff.

LOWER LUDLOW SHALES OF BIRMINGHAM
(64)

Present-day descriptions of this region are not available, so the old terminology must be used. Butler (1939, p. 55) reported *Monograptus flemingi* var. *delta* from the very base of the Lower Ludlow Shales of the Birmingham region; this graptolite is normally associated with the upper Wenlock, but Butler quoted Elles (1900, p. 403) to demonstrate that it is occasionally found in the Lower Ludlow. In the opinion of Rickards, this is highly unlikely. King and Lewis (1912, p. 439–440) mentioned that the contact of the Lower Ludlow Shales with the Aymestry Limestone was visible in the Netherton anticline, Birmingham.

LOWER NANTGLYN FLAGS GROUP
(34)

P. T. Warren (1967, written commun.) and his colleagues of the Institute of Geological Sciences of Great Britain have been revising the record of the northwest Denbighshire area and have collected graptolite assemblages representing the zones of *Cyrtograptus rigidus* (upper part) through *Monograptus ludensis*.

LOWER OWENDUFF GROUP
(4, 5)

This group has recently been redefined to include the Lough Mask, Kilbride, and Tonalee Formations (Laird and McKerrow, 1970). *Eocoelia curtisi* occurs in the Kilbride Formation and indicates a C_4-C_5 correlation. The Upper Owenduff Group contains lower Wenlock graptolites; therefore, it is concluded that the Lower Owenduff Group is of latest Llandovery age.

This name was used by Dewey (1963, p. 334) for a thin and impersistent unit that rests unconformably on the Ordovician in County Mayo and contains poorly preserved brachiopods. These fossils have been examined by us and are poorly preserved indeed. About all one could say is that they are definitely brachiopods.

In Galway, the Lower Owenduff Group is overlain by the Gowlaun Member of the Lettergesh Formation, and in Mayo by a similar unit, the Cregganbaun Conglomerate. The Mayo occurrences of the Lower Owenduff Group are therefore likely to be of the same late Llandovery age as those in Galway.

LOWER PHOSPHATIZED PEBBLE BED
(67)

Lawson (1954, p. 231) described this 1-centimeter band of phosphatized pebbles of siltstone and mudstone. It occurs between the Lower Siltstones and Upper Siltstones of Gorsley. Lawson (p. 233) considered that this bed may mark an important submarine disconformity involving a long period of interrupted deposition.

LOWER RED DOWNTON FORMATION
(63)

See Red Downton Formation

LOWER ROMAN CAMP BEDS
(48, 49, 50, 51, 52)

These beds have yielded *Monograptus leintwardinensis* in the northern part of the Llandovery area and *Neobeyrichia lauensis* in the Llangadock section, indicating a Leintwardinian age (Potter and Price, 1965, p. 385, 391). These beds cannot be distinguished from the Upper Cwm Clŷd Beds of the Llandeilo section, and both here and at Llangadock they are overlain unconformably by the Long Quarry Beds of Downton age. Elsewhere, they are followed in sequence by the Upper Roman Camp Beds.

LOWER SILTSTONES OF GORSLEY
(67)

These beds have yielded *Monograptus leintwardinensis* and *Dayia navicula* (Lawson, 1954, p. 231) and have been correlated with the Lower Leintwardine Beds of the type area by Holland and others (1963, p. 148-149). Lawson described this thin sequence as resting on the irregular top surface of the Gorsley Limestones, which he correlated with the Wenlock Limestone. Assuming this to be true, the unconformity between these two units must represent a considerable length of time.

LOWER SKELGILL BEDS
(29)

See Stockdale Shales.

LOWER SLEAVES OAK BEDS
(66)

Holland and others, (1963, Table II) correlated these beds with the Lower Bringewood Beds of the type area, apparently on general lithological and faunal similarity, because the units are transitional in nature to the Upper Sleaves Oak Beds and the Upper Bringewood Beds, respectively, that is, the Aymestry Limestone referred to by former workers. Squirrell and Tucker (1960) reported the graptolite *Monograptus chimaera* from these beds and compared them with the Upper Flaxley Beds of May Hill and the Upper Forest Beds of Usk.

LOWER TRAP OF TORTWORTH
(71)

The origin of the Lower Trap is uncertain; Reynolds (1924, p. 108) believed that it was intrusive, but Curtis (1955b, p. 5) stated that it was probably an extrusive lava. However that may be, the Lower Trap lies between the Cambrian Micklewood Beds and the C_5 Damery Beds in most parts of the Tortworth inlier, but it pinches out and is not present in the extreme northern parts of the inlier.

LOWER UNDERBARROW FLAGS
(29)

These beds contain shelly fossils characteristic of the Upper Leintwardine Beds of the Ludlow district (Shaw, 1971a, p. 370). Shaw (1971b, p. 606-607) collected the *Neobeyrichia scissa-N. lauensis* ostracode association from these beds. This association was originally recognized by Martinsson (1967, p. 370-373) at Gotland; he also collected it from the Upper Leintwardine Beds at Leintwardine, near Ludlow.

"LOWER WENLOCK" OF LLANGADOCK AND THE RIVER SEFIN
(49, 50)

Williams (1953, p. 198-199) has mapped these beds as resting unconformably on uppermost Llandovery, Upper Bala, and Llanvirn beds and has collected from them a fauna consisting of *Monograptus dubius, M. priodon, M. riccartonensis,* and *M. vomerinus,* as well as a rich shelly fauna. He collected various specimens of these species at several localities and concluded tentatively that they represented the zone of *M. riccartonensis* and that this was probably the lowest and only zone present in the lower half of the Wenlock of this area. Although there is a Wenlock overlap of the Llandovery in the Llandeilo area, there is no evidence of an unconformity between the Wenlock and the Llandovery at Llangadock (Cocks and others, 1971, p. 108).

LOWER WHITCLIFFE BEDS
(62, 63)

These beds, which have sometimes been referred to as the *Rhynchonella* Flags, have not yielded a graptolite fauna (Holland and others, 1959). Their position above beds with the *M. leintwardinensis* zonal assemblages would suggest a correlation with or above this zone.

LOWER WOOTON BEDS
(66)

Squirrell and Tucker (1960, p. 145, 157, 173) described these beds and reported the graptolites *Monograptus chimaera, M.* cf. *colonus* var. *compactus, M. tumescens,* and *M. varians,* which they correlated with the zones of *M. nilssoni* and *M. scanicus.* Holland and others (1963, p. 148) compared the base of the Lower Wooton Beds with the Lower Elton Beds of Ludlow, and pointed out that the presence of *M.* cf. *colonus* var. *compactus* and *M. varians* in the higher part of the division suggests a correlation of these beds with the Middle Elton Beds. Squirrell and Tucker (1960) compared the fauna and lithology of the Lower Wooton Beds with the Lower Flaxley Beds of May Hill and the basal part of the Lower Forest Beds of Usk.

LUDLOW BONE BED
(62, 63, 64, 65, 66, 69)

See Grey Downton Formation.

"LUDLOW SERIES" OF FRESHWATER EAST AND FRESHWATER WEST
(47)

These beds have not yielded diagnostic fossils and their age is uncertain. They are overlain disconformably by the Red Marls, the basal unit of the lower Old Red Sandstone. At Freshwater West they rest on the Ordovician, and at Freshwater East they succeed the "Wenlock Series" of doubtful age. Sanzen-Baker (1972) reported Downton spores.

LYNE WATER FISH BEDS
(23)

Included under this heading are all the beds above the Red Igneous Conglomerate and beneath the unconformable lower Old Red Sandstone. The fish beds occur only in the Lyne Water area and have been dated by Westoll (1951, p. 7) as late Wenlock or early to middle Ludlow. In the North Esk area, a quartzitic conglomerate occurs in the probable equivalents of the fish beds. This conglomerate is thought to be equivalent to the Kirk Hill Conglomerate of Carmichael, the Hareshaw Conglomerate of the Hagshaw Hills, and the Middlefield Conglomerate of Lesmahagow.

MAIN CONTORTED GROUP
(56)

Earp (1940, p. 4) regarded this group as a local development of the *Monograptus leintwardinensis* shales and of the Wilsonia Grits, and he listed *Monograptus leintwardinensis* var. *incipiens* from the lower part of the group and *M. leintwardinensis* from the upper part.

MANSE MUDSTONES
(22)

See Carmichael Burn Group.

MAXWELLSTON MUDSTONES
(18)

These beds are the old Maximus Mudstones of Lapworth (1882) and contain graptolites of the subzone of *Rastrites maximus* (Cocks and Toghill, 1973).

"MIDDLE BIRKHILL" OF ABERGWESYN AND PUMPSAINT
(42)

Four zones, supported by long faunal lists, have been recognized in the "Middle Birkhill" of this area, the zones of *Monograptus triangulatus, Diplograptus (Mesograptus) magnus, Monograptus leptotheca,* and *M. convolutus* (Davies, 1933, p. 180, 184).

MIDDLE COLDWELL BEDS
(29)

See Upper Coniston Flags.

MIDDLE ELTON BEDS
(62, 63)

The Middle Elton Beds (Holland and others, 1959) are the Lower Ludlow Shales of the *M. nilssoni* and *M. scanicus* zones of Wood (1900, p. 425-428). They have been mapped in the Wenlock area, where they also contain *M. nilssoni-scanicus* assemblages (Shergold and Shirley, 1968, p. 122).

MIDDLEFIELD CONGLOMERATE
(20)

See Dungavel Group.

MIDDLE LLANDOVERY OF LLANDOVERY AND GARTH
(51, 52, 53)

Jones (1925b, p. 361-365) recognized three units, B_1, B_2, and B_3 in the River Ydw section; in the northern Llandovery area (1949, p. 51-53) he also recognized three units, B_a, B_b, and B_c. Lithologies are variable between the areas, and Jones did not wish to imply any correlation of these two sequences. He mapped the middle Llandovery beds as resting unconformably on various units of the lower Llandovery sequence.

Graptolites have been found in units B_a and B_c in the northern area. Jones lists *Climacograptus scalaris* from B_a and *Monograptus decipiens, M.* of the *lobiferus* group, *M. regularis, Orthograptus cyperoides,* and *Climacograptus scalaris* from B_c. These occurrences suggest a lower limit of the zone of *Monograptus gregarius* for B_a and, according to Jones, a correlation with the zone of *Cephalograptus cometa* for B_c. The *C. cometa* zone is equivalent to the upper part of the *Monograptus convolutus* of the present usage.

At Garth, Andrew (1925, p. 403) mapped a small development of mudstone as B or middle Llandovery and matched these beds lithologically with unit B_3 of Llandovery. He collected *Monograptus decipiens* from unit B; this species ranges from the *M. gregarius* zone to the *M. sedgwicki* zone.

Williams (1951, p. 101) reported *Stricklandia lens intermedia* from the middle Llandovery of the northern Llandovery district.

MIDDLE LLANGIBBY BEDS
(69)

Walmsley (1959, p. 510) matched this unit with the Whitcliffe Flags of the type area because of lithologic and faunal similarities; he also matched it with the Upper Longhope Beds of May Hill. Holland and others (1963, Table II) correlated the Middle and Upper Llangibby Beds with the Whitcliffian of Ludlow.

MIDDLE SKELGILL BEDS
(29)

See Stockdale Shales.

MILL COVE FORMATION
(1)

See Dunquin Group.

MILLIN "STAGE"
(44)

Strahan and others (1914, p. 81) recognized two units in this stage, the Uzmaston Beds and the Canaston Beds. The Uzmaston Beds follow the Haverford "Stage" with apparent conformity; the lower part of the unit is unfossiliferous, but *Eospirifer radiatus* and *Cyrtia exporrecta* occur at about the middle of the sequence, indicating a C_3 or later age for this part of the Uzmaston Beds. Unfortunately, the sequence of the Canaston Beds on the Uzmaston Beds cannot be demonstrated in the field, but very few strata can actually be missing because the Canaston Beds have yielded a rich C_6 fauna, including *Costistricklandia lirata* and *Palaeocyclus* sp. (Strahan and others p. 109). The top of the Canaston Beds is obscured by the lower Old Red Sandstone, which rests on them with angular unconformity.

MINSTERLEY FORMATION
(60)

Ziegler and others (1968b, p. 744) have applied the name Minsterley Formation to strata of the Minsterley area that Whittard (1932, p. 877) called Purple Shales (Hughley Shales). Whittard reported *Climacograptus* sp., *Monograptus halli*, and *M. becki* from these beds, a fauna that points to the *M. turriculatus* zone. The upper extent of the formation in time is unknown, but the underlying Venus Bank Formation is probably of middle or early late Llandovery age.

MOCKTREE SHALE MEMBER OF MALVERN
(65)

See Upper Ludlow Formation of Malvern.

MOELFRE GROUP
(39)

See Llanidloes "Stage."

MONOGRAPTUS BEDS
(39)

See Clywedog "Stage."

MONOGRAPTUS LEINTWARDINENSIS SHALES
(56)

Earp (1938, p. 132) described this unit, and it is assigned to the zone of *M. leintwardinensis* because of the common occurrence of this species.

MONUMENT FORMATION
(20)

See Waterhead Group.

MOUGHTON WHETSTONES
(30)

The list reported by King and Wilcockson (1934, p. 23) as "indicating a proximity to the junction of the *C. lundgreni* and *M. nilssoni* zones" is thought to contain specimens from rocks of both this and the overlying formation. One of us (Rickards) has collected a *C. lundgreni* zone fauna from rocks of the Moughton Whetstone lithology. The original collection has a *C. lundgreni* zone fauna on slabs with the Moughton Whetstones lithology and a *P. nilssoni* zone assemblage on slabs with a Horton Flags lithology. The locality has been revised by McCabe (1972).

MOYLUSSA GROUP
(3)

The Moylussa Group is sparingly fossiliferous but overlies the Craglea Group of late Wenlock–early Ludlow age and is overlain with angular unconformity by the Old Red Sandstone (Weir, 1962, p. 248).

MUDSTONE AND LIMESTONE FROM THE WARE BOREHOLE
(79)

Whitaker and Jukes-Browne (1894, p. 506) have given a complete description of this borehole, which passed through 0.6 meters of shelly limestone and then 10 meters of mudstone before stopping. The limestone is unconformably overlain by Lower Greensand. Previous to this description, Hopkinson (1880, p. 247) published a long faunal list containing such common Wenlock forms as *Meristella tumida, Cyrtia exporrecta, Howellella "elevata," Eospirifer plicatella, Rhynochonella cuneata, Gypidula galeatus, Clorinda linguifera, Strophonella euglypha, S. depressa,* and *Pentlandina antiquata.* Etheridge (1881, p. 230) claimed that the entire list of fossils could be matched with the Wenlock of Wenlock Edge and Dudley. Cocks and others (1971, p. 125) have identified *Meristina obtusa* from this material and regard it as late Wenlock in age.

MUDSTONES AND SANDSTONES OF CARDIFF
(70)

See Silurian of Cardiff.

MUDSTONES AND SLUMPED SILTSTONES OF THE
WEELEY BOREHOLE
(80)

Whitaker and Thresh (1916, p. 344) described this borehole, which entered the "?Silurian or Cambrian?" beneath an unconformity with the Gault and penetrated these older rocks to a depth of 38 meters. Cocks and others (1971, p. 125) report that T. R. Lister has identified microfossils, indicating a Llandovery age for these beds.

MUDSTONES OF LISBELLAW
(11)

These mudstones occur both above and below the Lisbellaw Conglomerate. Harper and Hartley (1938, p. 75) list graptolites from three localities in the lower mudstones, all of which they believed represented the zone of *Monograptus gregarius*. The first list contains *M. gregarius, M. incommodus, M. revolutus*(?), and *Climacograptus rectangularis*. The second assemblage came from the top of these beds, just beneath the Lisbellaw Conglomerate: *M. gregarius, M. regularis, M. incommodus, M. argutus, M. sandersoni, M. triangulatus, Rastrites peregrinus*, and *Climacograptus rectangularis*. A third locality yielded *Monograptus incommodus, Climacograptus innotatus*, and *C. hughesi*(?). The base of the series is unknown.

Graptolites have been reported from two localities in the upper mudstones by Harper and Hartley (1938, p. 75-76). The first yielded *Petalograptus* cf. *folium, Monograptus leptotheca*, and *Diplograptus foliaceus*, and the second, *Monograptus gemmatus*. These probably represent the zone of *M. gregarius* or *convolutus* as do graptolites from the underlying Lisbellaw Conglomerate and lower mudstones.

MUDSTONES OF THE EASTERN MENDIPS
(72)

Reed identified several collections of shelly fossils from these beds for Reynolds (1912, p. 78-79) and was of the opinion that the fossils were of Wenlock age. His listing of "*Spirifer*" *sulcatus* and "*Spirifer*" *elevatus* certainly indicates a post-C_2 age for the beds. Reynolds (p. 80) concluded on mapping evidence that the mudstones rest above the pyroxene andesite flow rock. The upper Old Red Sandstone rests with pronounced angular unconformity on the mudstones, according to Kellaway and Welch (1948, p. 16).

MUDSTONES OF THE LOWESTOFT BOREHOLE
(84)

Strahan (1913b, p. 87-88) described this borehole, which penetrated into the "Cambrian or Ordovician" 62 meters beneath the unconformity with the Lower Greensand. Stubblefield re-examined the material for Bullard and others (1940, p. 88) and reported *Lingula, Orbiculoidea,* and *Ceratiocaris?*; all of these were found to be fragmented. These fossils could indicate either Ordovician or Silurian strata, but Cocks and others (1971, p. 125) reported that T. R. Lister identified Upper Silurian spore assemblages from these beds.

MUDSTONES OF THE SHALFORD BOREHOLE
(74)

A description of this borehole, which entered 26 meters of Silurian rock beneath the angular unconformity with the Carboniferous, may be found in Falcon and Kent (1960, p. 47-48). Smith (1959, p. 48) reported *Monograptus* cf. *distans,* identified by Bulman, which would indicate a correlation with the *Monograptus convolutus* or *M. sedgwicki* zones. Ziegler has examined the brachiopod material preserved in the Geological Survey Museum in London and has discovered *Stricklandia lens* cf. *progressa* and *Eocoelia hemisphaerica* or *E. intermedia.* The brachiopod and graptolite evidence taken together would indicate a C_1-C_2 date for these rocks.

MUDSTONES WITH GREYWACKES OF THE STUTTON BOREHOLE
(82)

A total of 162 meters of this rock was penetrated beneath the unconformity with the Gault, and an orthocone nautiloid was reported (Whitaker, 1906, p. 140). Stubblefield re-examined the material (Bullard and others, 1940, p. 87) and recorded a new find: ". . . appears to be a eurypteridian, possibly a leg fragment; it shows surface ornament, which somewhat resembles that seen on the legs of *Carcinosoma punctatus* (Salter)." Cocks and others (1971, p. 125) reported that T. R. Lister has identified microfossils from these beds, which indicate an age of early Wenlock or perhaps later Llandovery.

MULLAGHNABUOYAH BEDS
(12)

See Little River Group.

MULLOCH HILL FORMATION
(17)

The Mulloch Hill Formation is probably of early Llandovery age because it rests beneath the Glenwells Shale, which includes a *M. cyphus* zone assemblage, and some distance above the Drummuck Group, which contains the *Dicellograptus anceps* zone, the highest graptolite zone of the Ordovician (Lamont, 1935, p. 301).

MYHERIN GROUP
(37)

See Ystwyth "Stage."

NANT-Y-BACHE GROUP
(35)

Wills and Smith (1922, p. 202) implied that the Nant-y-bache Group included at least parts of the zones of *Monograptus nilssoni, M. scanicus,* and ?*M. tumescens,* but they list only one distinctive assemblage, which they interpreted as the zone of *M. scanicus.* It contains *M. chimaera, M. chimaera* var. *salweyi, M. colonus, M. dubius,* and *M. variens,* a fauna which cannot strictly be referred to the *M. scanicus* zone.

NANT-YSGOLLON SHALES
(38)

Wood (1906, Table 1) referred these beds to the basal Wenlock zones of *Cyrtograptus murchisoni* and *Monograptus riccartonensis,* but only the former zone is represented by a large fauna. Wood identified the zone of *M. riccartonensis* solely on the fact that this species becomes more abundant above the *Cyrtograptus murchisoni* fauna. In other regions, *M. riccartonensis* is extremely rare, if present at all, below the *M. riccartonensis* zone.

NASH SCAR AND DOLYHIR LIMESTONES
(54)

These limestones, which crop out a few kilometers apart, were considered to be the same age by Kirk (1951a, p. 56). In the discussion of Kirk's paper (p. 57), the occurrence of *Stricklandia lirata* in the base of the Dolyhir Limestone was mentioned, indicating a late Llandovery age. An upper limit for the limestone is indicated by the occurrence of *Cyrtograptus symmetricus* (= *rigidus*) in the overlying Wenlock Shales. This graptolite does range from its own zone into the overlying zone of *C. linnarssoni* but is less common there.

The Nash Scar Limestone rests on lower upper Llandovery sandstone, but the Dolyhir Limestone rests directly on Precambrian conglomerate and mudstone.

NEWLANDS FORMATION
(17)

This new formation was proposed by Cocks and Toghill (1973; see also Cocks, 1971a, p. 225) who collected the middle Llandovery brachiopod *Stricklandia lens intermedia* from it. It replaces in part the old Newlands Sandstone of Lapworth (1882).

NEWSIDE ARENITE
(22)

Diagnostic fossils have not yet been found in this unit, but Rolfe (1960, p. 251) implied a lithologic correlation with the Glenbuck Group of the Hagshaw Hills inlier. Both of these units are underlain by Igneous Conglomerate and overlain by Quartzite Conglomerate, which makes this correlation likely.

OLDCHAPEL MUDSTONES
(39)

See Llanidloes "Stage."

"OLD RED SANDSTONE?" OF BUCKNELL
(55)

These rocks, which were mapped by Stamp (1919), have been correlated with the Red Downtonian Beds of surrounding areas by Holland (1959b, p. 473).

ORANGE-WEATHERING SHALES OF LLANGRANOG AND ABERYSTWYTH
(43)

Hendriks (1926, p. 130-132) has identified the zones of *Glyptograptus persculptus, Cephalograptus* (= *Akidograptus*) *acuminatus,* and *Monograptus cyphus* from these shales.

ORTHONOTA MUDSTONES
(53)

Straw (1937, p. 450) lists *Monograptus leintwardinensis* from the lower part of these beds, indicating a correlation with the zone of *M. leintwardinensis.* Beneath the Orthonota Mudstones, the highest beds of the Wilsonia Shales also have yielded this species.

OUGHTY GROUP
(5)

See Croagh Patrick "Series."

OWENWEE GROUP
(5)

See Croagh Patrick "Series."

PALE MUDSTONES OF LLANGRANOG AND ABERYSTWYTH
(43)

Hendriks (1926, p. 122, 123) believed that these Pale Mudstones extended from the zone of *M. convolutus* to *M. turriculatus,* but she listed only one fauna and correlated this with the zone of *Cephalograptus cometa,* that is, the upper part of the zone of *Monograptus convolutus.* Jones (*in* Wood and Smith, 1959, p. 191) has since stated that the overlying Aberystwyth Grits belong to the lowest part of the *M. turriculatus* zone in this area.

PALE SHALES
(39)

See Llanidloes "Stage."

PARISHHOLM CONGLOMERATE
(21)

This conglomerate (Rolfe, 1962, p. 255) is otherwise known as the "Igneous Conglomerate," a widespread lithologic unit that is recognized in the Carmichael inlier as the Fence Conglomerate, and in the North Esk inlier as the Red Igneous Conglomerate. In the North Esk inlier, this unit overlies beds of probable early Wenlock age, and in all three inliers it underlies fish-bearing beds of possible late Wenlock to middle Ludlow age (Westoll, 1951, p. 7). Pebbles containing stromatoporoid and bryozoan fragments have recently been collected from this conglomerate. These forms are thought to be of Wenlock age (Rolfe and Fritz, 1966, p. 162); therefore, the Parishholm Conglomerate and its correlatives would be later, at least, than some part of the Wenlock. Rolfe and Fritz concluded that a paraconformity exists beneath these conglomerates and that they are probably of late Wenlock or Ludlow age.

PASSAGE BEDS
(57)

This is a relatively thin sequence, which occurs between the upper Llandovery Horizon V_3 of *Monograptus turriculatus* zone of *M. crenulata* zone age and the Salopian, which contains the zone of *Cyrtograptus murchisoni* in its base (King, 1928, p. 691). King collected *Monograptus crenulatus* from the Passage Beds at one locality, which would suggest that they belong to the upper Llandovery.

PASSAGE FORMATION
(20)

See Waterhead Group.

PATRICK BURN FORMATION
(20)

See Priesthill Group.

PEGWN MUDSTONES AND GRITS
(39)

This unit has yielded fossils belonging to the zone of *Cyrtograptus murchisoni* (Jones, 1945, p. 325). Its upper limit is not exposed within the Llanidloes area.

PENCHRISE BURN BEDS
(26)

See Riccarton Group.

PENCLEUCH SHALE
(17, 18)

The Pencleuch Shale is Lapworth's (1882) old Sedgwicki Shale, but Cocks and Toghill (1973) have collected from the upper beds a large graptolite fauna representative of the convolutus zone, including *Monograptus convolutus* itself.

PENKILL FORMATION
(18)

Cocks and Toghill (1973) reported that these beds contain a *M. turriculatus* zonal assemblage through most of their thickness, but they have also yielded a *M. crispus* zonal assemblage from the highest horizons.

PENMACHNO BEDS
(31)

This unit has yielded some specimens of *Monograptus scanicus* and is positioned above beds bearing *M. nilssoni* and below beds with a good *M. scanicus* assemblage (Woods and Crosfield, 1925, p. 177); therefore, it has been assigned to the zone of *M. scanicus*.

PENTAMERUS BEDS
(62)

Whittard (1928, p. 746) has reported the following graptolites from the Pentamerus Beds: *Climacograptus scalaris, Glyptograptus* aff. *tamariscus, Monograptus becki,*

M. gemmatus, and *M. nudus*. Several of these species were listed from various localities, but they all occur together at one locality, with the exception of the *Climacograptus scalaris*. All of these species are common to the *M. sedgwicki* zone, except *M. becki*, which is known from the two overlying zones. A late Llandovery age is also suggested by the occurrence of the brachiopod species, *Eocoelia hemisphaerica* and *E. intermedia*, and *Stricklandia lens progressa* and *S. lens ultima* (Ziegler and others, 1968b, p. 746, 748). These species were not found together but came from various sections and suggest a span of C_1 or C_2 to C_3 or C_4. Cocks and Rickards (1969, p. 217) have recently discovered the *M. convolutus* and *M. sedgwicki* zones in a borehole near Church Stretton; therefore, middle Llandovery horizons are represented as well.

PEN-Y-GLOG GROUP
(35)

Three graptolite zones have been identified in this unit: *Cyrtograptus murchisoni*, *Monograptus riccartonensis*, and *Cyrtograptus linnarssoni* (Wills and Smith, 1922, p. 201-203; Elles, 1900, p. 399). Thus, the lower Wenlock is well represented by fossils, and in fact, the underlying Ty-draw Slates contain the top Llandovery graptolite zone. The Pen-y-glog Group probably spans all the Wenlock, because three forms, *M. dubius*, *M. flemingii* var. *gamma*, and *M. flemingii* var. *delta*, indicative of the zone of *Cyrtograptus lundgreni*, have been reported (Elles, 1900). It may be mentioned that the Denbigh Grit, known locally as the Pen-y-glog Grit, occurs within the Pen-y-glog Group but only in the western part of the area where it overlies the *Cyrtograptus linnarssoni* fauna.

PERSCULPTUS MUDSTONES
(39)

See Clywedog "Stage."

PERTON BEDS
(66)

Squirrell and Tucker (1960, p. 160) correlated these beds with the Whitcliff Flags (Whitcliffian Stage) of Ludlow, the Upper Longhope Beds of May Hill, and the Middle and Upper Llangibby Beds of Usk. They occur above the Upper Bodenham Beds of *Monograptus leintwardinensis* zone age and below the Rushall Beds of Downton age.

PHOSPHATIZED PEBBLE BED
(68)

This bed is the local equivalent of the Ludlow Bone Bed (Lawson, 1954, p. 233; 1955, p. 107).

PLAS-NEWYDD BEDS
(31)

Woods and Crosfield (1925, p. 178) have collected a *Monograptus scanicus* assemblage from these beds.

PLATYSCHISMA HELICITES BEDS
(54, 56)

These beds have been described by Earp (1938, p. 135-137) and Holland (1959b, p. 462-463). They are the lowest beds in the area containing a fauna of the Downton type (brackish marine), and on this basis they have been assigned to the base of the Downton.

PLEWLANDS SANDSTONE
(20)

See Dungavel Group.

PLOXGREEN FORMATION
(60)

These beds were named and briefly described by Palmer (1970b, p. 342). Cocks and Rickards (1969, p. 224) reinterpreted the graptolite list of Whittard (1932) and concluded that middle Wenlock horizons are present in this area.

POLLANOUGHTY GROUP
(5)

See Croagh Patrick "Series."

POLLAPHUCA FORMATION
(15)

See Shales, Grits, and Conglomerates of Slieve Bloom, Knockshigowna, Blessington, and Chair of Kildare.

PONT ERWYD "STAGE"
(36, 37)

O. T. Jones (1909, p. 475-508) originally defined this stage in the district around Plynlimon and Pont Erwyd. Three groups of strata constitute the Pont Erwyd "Stage" in this district; the Eisteddfa Group at the base, followed by the Rheidol Group, and the Castell Group at the top. In the Eisteddfa Group, Jones (p. 504) recognized two zones: the basal zone of *Glyptograptus persculptus*, which rests on the Plynlimon "Stage" of highest Ordovician age, as shown by the zone of *Dicellograptus anceps,* and the zone of *Akidograptus acuminatus.* The Eisteddfa Group is therefore of basal Llandovery age. Jones recognized four zones in the Rheidol Group: the zones of *Monograptus atavus, M. rheidolensis, M. cyphus,* and *M. communis.* The zone of *M. communis* at the top of the Rheidol Group is of local character, but it contains *M. gregarius* and is probably equivalent to the zone of *M. gregarius.* Finally, the Castell Group contains the zones of *M. convolutus, Cephalograptus cometa,* and *Monograptus sedgwicki.*

The Pont Erwyd "Stage" has also been recognized in the Machynlleth district, but here the constituent units are the Cwmere Group at the base, and the Derwen Group at the top (Jones and Pugh, 1915, p. 349). Three faunal units were recognized in the Cwmere Group: the zones of *Glyptograptus persculptus* and *Akidograptus acuminatus* and the *Monograptus* spp. beds. The first two units allow a direct correlation of the base of the Cwmere Group with the basal Llandovery Eisteddfa Group of Pont Erwyd. The *Monograptus* spp. beds have yielded faunas from several horizons, including forms such as *Monograptus triangulatus* and *M. fimbriatus,* which indicate that the Cwmere Group extends at least into the zone of *M. gregarius.* Jones and Pugh (1915, p. 380) recognized five zones in the overlying Derwen Group. In stratigraphical order, they are the zones of *Mesograptus magnus, Monograptus leptotheca, M. regularis, M. sedgwicki,* and *M. halli.* They correlated the first zone with the top of the *M. gregarius* zone of Moffat, the second two with the zones of *M. convolutus* and *Cephalograptus cometa* of Pont Erwyd, and the top two with the *Monograptus sedgwicki* zone of Pont Erwyd.

POWIS CASTLE GROUP
(58)

Wade (1911, p. 431–433) described these beds as resting unconformably on various horizons of the Upper Ordovician. They do not contain a diagnostic fauna but may be dated fairly accurately by a consideration of the underlying and overlying stratigraphic units. The highest Ordovician rocks of the area, the Gwern-y-brain Group, contain a fauna that Wade believed to be transitional between the Ordovician and Silurian; he reported *Trinucleus* cf. *seticornis, Diplograptus modestus* cf. var. *parvulus, Orthograptus truncatus* cf. var. *socialis,* and *Glyptograptus persculptus.* The graptolite fauna is indicative of the *G. persculptus* zone and is usually taken to indicate the base of the Llandovery, but actually all the forms, with the exception of *Diplograptus modestus* var. *parvulus,* are known to range into the Ordovician. The Powis Castle Group, then, is probably post-Ordovician, and the overlying Cloddiau Group has yielded graptolites known to range throughout the lower Llandovery, placing an upper limit of late lower Llandovery.

PRIESTHILL GROUP
(20)

This group contains five formations that are, from base to top, the Patrick Burn, Castle, Kip Burn, Blaeberry, and Dunside Formations. Fossils of correlative value are lacking, but Walton (1965, p. 196) implied a lithologic correlation of the Patrick Burn Formation with the Ree Burn Formation of the Hagshaw Hills inlier, and Rolfe and Fritz (1966, p. 162) concluded that the upper four formations are missing in the Hagshaw Hills because of an unconformity.

PROTOVIRGULARIA GRITS
(18)

Cocks and Toghill (1973) have collected both *M. crispus* and *M. griestoniensis* zonal faunas from these beds.

PYROXENE ANDESITES OF THE EASTERN MENDIPS
(72)

This volcanic unit occurs between two units of possible Wenlock age (Reynolds, 1912, p. 82).

PYSGOTWR GRITS
(42)

See "Gala" of Abergwesyn and Pumpsaint.

QUARRY ARENITE
(21)

The age of the Quarry Arenite is uncertain (Rolfe, 1962, p. 261); it could range as high as the Ditton because the overlying lower Old Red Sandstone is not precisely dated (Rolfe, 1960, p. 253).

RED DOWNTON BEDS
(64)

See Red Downton Formation.

RED DOWNTON FORMATION
(62, 63, 64, 65)

The definition of the top of the Red Downton Formation, which constitutes the upper half of the Downton, has been a matter of dispute. Some have put the *Psammosteus* Limestones and the *Traquairaspis* fish zones in the Downton (White, 1950, p. 53), while others have drawn the line beneath these horizons (Allen and Tarlo, 1963, p. 130). The latter course seems to be more generally accepted because the Downton-Ditton contact would correspond more closely to the Silurian-Devonian boundary as understood on the Continent (Holland, 1965, p. 215).

Dineley and Gossage (1959, p. 222) recognized the Holdgate Sandstones between the Lower and Upper Red Downton Formations at Cleobury Mortimer, and Phipps and Reeve (1967, p. 353) applied the name Ledbury Group to the corresponding horizons at Malvern. Ball and White (1961, p. 180) described these rocks at Brown Clee Hill, and Ball (1951, p. 226) described them at Birmingham.

RED DOWNTONIAN BEDS
(54, 56, 59)

The distinctive fish faunas are lacking in these rocks (Earp, 1938, p. 138-140; Holland, 1959b, p. 462-464), but similar Red Marls along the strike to the southwest yielded lower Ditton fish and have been correlated with the late Downton and early Ditton. Holland (1959b, p.472) expressed the view that only upper Downton beds are present in the Clun Forest area, the higher beds being removed by recent erosion. This unit has been reported from the Long Mountain area (Cocks and others, 1971, p. 110), but the top has been removed by erosion.

RED IGNEOUS CONGLOMERATE
(23)

This conglomerate is unfossiliferous (Mitchell and Mykura, 1962, p. 12), but it is probably of Wenlock age because it overlies beds of lower Wenlock age and underlies fish beds of late Wenlock or early to middle Ludlow age (Westoll, 1951, p. 7). This unit has been correlated with the Parishholm and Fence Conglomerates, which are thought to be of late Wenlock age (Rolfe and Fritz, 1966, p. 165).

RED MARL GROUP OF CARDIFF
(70)

Heard and Davies (1924, p. 494) described these rocks; they follow rocks presumed to be of Ludlow age without apparent stratigraphic break and contain early Devonian pteraspidian fish at one horizon.

RED MARLS OF LLANDEILO, LLANGADOCK, LLANDOVERY, AND BUILTH
(48, 49, 50, 51, 52, 53)

These rocks were described by Potter and Price (1965, p. 392), who reported the Lower Devonian fish, *Pteraspis leathensis*, from a loose block that was probably derived from the higher beds of the unit. The lower beds contain the brackish water community typical of the Downton of the Welsh Borderland,

and Potter and Price concluded that the Red Marls are Late Silurian and Early Devonian in age.

RED MARLS OF PEMBROKESHIRE
(44, 45, 46, 47)

These marls rest with profound angular unconformity on many groups of older rocks at Haverfordwest and Rosemarket (Strahan and others, 1914, p. 113) and with disconformity on Silurian and Ordovician rocks at Freshwater East and Freshwater West (Dixon, 1921, p. 27), whereas at Marloes and Winsle there appears to be a conformable passage from the Sandstone Series of presumed Ludlow age (Cantrill and others, 1916, p. 89).

All the authors mentioned above reported pteraspid fish remains from the higher beds of the Red Marls. These fish are usually taken to indicate a Lower Devonian age, but their lower stratigraphic limit has never been established. Richardson and Lister (1969, p. 211, 212, 252) recorded the spore *Synorisporites verrucatus* from faulted Red Marls at Freshwater East. Part of this formation is therefore apparently of Downton age.

RED MARLS OF WOOLHOPE, GORSLEY, MAY HILL, AND USK
(66, 67, 68, 69)

These rocks have been mapped by Squirrell and Tucker (1960), Lawson (1955), and Walmsley (1959) and are the local lithologic equivalents of the Red Downton Formation of Ludlow.

RED SANDSTONES OF THE CURLEW MOUNTAINS
(7)

The Red Sandstones are unfossiliferous, but they have been correlated with the red Basal Beds (Lough Mask Formation) of northwest County Galway by Charlesworth (1960, p. 48) on lithologic similarities.

REE BURN FORMATION
(21)

See Hagshaw Group.

RESERVOIR BEDS AND GUTTERFORD BURN FLAGSTONES
(23)

Mitchell and Mykura (1962, p. 12) have included the Gutterford Burn Flagstones in this unit. They report *Palaeocyclus porpita, Monograptus* cf. *priodon,* and *M.*

vomerinus aff. *gracilis.* The *Palaeocyclus* indicates a C_6 age, and the graptolites are in accord with this age. The base of the Reservoir Beds is unknown.

RHAYADER PALE SHALES
(40, 41)

This unit is largely unfossiliferous and its top is not seen within the area. In Caban-coch it rests on the Caban Group, which contains the zone of *Monograptus sedgwicki.* The Rhayader Pale Shales contain the zone of *Monograptus sedgwicki* in the Rhayader area (H. Lapworth, 1900, p. 82).

RHEIDOL GROUP
(37)

See Pont Erwyd "Stage."

RHIWISG BEDS
(31)

These beds were named by Woods and Crosfield (1925, p. 175), who collected a *Monograptus nilssoni* zonal assemblage from them.

RHUDDNANT GROUP
(37)

See Ystwyth "Stage."

RHYMNEY GRIT
(70)

See Silurian of Cardiff.

RHYNCHONELLA BEDS OF BUCKNELL
(55)

These beds were described by Stamp (1919, p. 228) and have not yielded a graptolite fauna. Holland (1959b, p. 469-472) discussed the correlation of these beds and those of the adjacent Knighton area and concluded that the Rhynchonella Beds were equivalent to the Upper Knucklas Castle Beds and the Wern Quarry Beds. Holland and others (1963, Table II) assigned the Rhynchonella Beds to the lower Whitcliffian.

RICCARTON AND RAEBERRY CASTLE BEDS
(27)

See Riccarton Group.

RICCARTON GROUP
(26, 27)

Warren (1964, p. 196) has defined four divisions of the Riccarton Group on the basis of abundant graptolite faunas. The lowest unit, the Stobs Castle Beds, contains the zone of *Cyrtograptus murchisoni*. The next unit, the Shankend Beds, contains a fauna that Warren (p. 201) considered intermediate between the zones of *C. murchisoni* and *Monograptus riccartonensis*. The third unit, the Penchrise Burn Beds, is characterized by the zone of *M. riccartonensis*. At the top of the sequence are the Caddroun Burn Beds, which yielded faunas representing both the zones of *Cyrtograptus symmetricus* (= zone of *C. rigidus*) and *C. lundgreni*. The intermediate zones are not well represented, but Warren concluded (p. 205) that the sequence was complete and that the apparent zonal gap was due to lack of exposures and absence of diagnostic zonal forms.

Warren (1964, p. 208) questioned whether the poorly fossiliferous Hawick Rocks were the lateral equivalent of the Riccarton Group, partly on structural grounds, but Rust, in the discussion of Warren's paper, expressed doubt as to the structural evidence. Later, Rust (1965, p. 105) produced graptolitic evidence from both the Hawick Rocks and Riccarton Beds of the Whithorn area to suggest that the former are upper Llandovery and that the latter are basal Wenlock. The Riccarton Beds of the Whithorn have yielded *Cyrtograptus* aff. *insectus*, *C. pulchellus*, *Monograptus vomerinus* var. *basilicus*, *M. priodon*, and *Retiolites geinitzianus* (Rust, 1965), which indicates a correlation with the Stobs Castle Beds.

The Raeberry Castle Beds were thought (Peach and Horne, 1899, p. 80) to lie above the Riccarton Beds of Kirkcudbright, but this is now uncertain (Walton, 1965, p. 185).

RILBURY SILTSTONE
(65)

See Aymestry Formation of the Malvern District.

ROSEMARKET "STAGE"
(45)

The beds of this unit rest, with basal conglomerates, on volcanic rocks of the Benton Series and possibly also on the presumably older plutonic rocks of the Johnston Series (Strahan and others, 1914, p. 109-112). The Ashgillian Slade Beds recently have been discovered in this area (Cocks and others, 1971, p. 108). Pebbles of the Benton Series are dominant in the basal conglomerates of the Rosemarket "Stage," while pebbles possibly representing the Johnston Series are rare and questionable. The Benton Series and the Johnston Series have been dated isotopically as late Precambrian (S. Moorbath, 1969, oral commun.).

Pentamerus oblongus has been listed from most of the outcrops of the Rosemarket Stage (Strahan and others, 1914; Cantrill and others, 1916, p. 84-85), and forms identified as such have a range of B$_1$ to C$_5$ at the Llandovery type area. A graptolite specimen identified as cf. *Climacograptus scalaris* var. *normalis* has been collected from beds considerably above the base of the Rosemarket "Stage" (Cantrill and others, 1916; O. T. Jones, 1921, p. 159), which would suggest that this horizon is at least as old as the zone of *Monograptus gregarius*. Jones pointed out that if the specific and varietal identifications of this specimen were wrong, the minimum age would be the zone of *M. turriculatus*. One of us (Ziegler) has collected *Eocoelia hemisphaerica* from the same locality, so we assume that the graptolite identification is in error and that the age of this part of the Rosemarket "Stage" is early upper Llandovery.

RUBERY SANDSTONE
(64)

Wills and others (1925, p. 72-73) and Wills and Laurie (1938, p. 179-180) have provided descriptions and faunal lists of this unit. They report *Stricklandia lirata forma alpha*, *Coelospira hemisphaerica*, *Pentamerus oblongus*, and *Palaeocyclus* sp. These identifications have been checked by Ziegler on material kept in the Department of Geology, University of Birmingham, England. The stricklandiid identification is correct and should be listed *Costistricklandia lirata* var. *alpha*; the *Coelospira* is *Eocoelia curtisi*; the *Pentamerus* is *Pentameroides* sp. These identifications, except the *Palaeocyclus*, indicate a C$_5$ age for the Rubery Sandstone, but even *Palaeocyclus* is occasionally found in beds of C$_5$ age although it is much more common in beds of C$_6$ age. The Rubery Sandstone is a relatively thin unit and rests, with basal conglomerate, on Cambrian quartzite.

RUBERY SHALES
(64)

The Rubery Shales have yielded an abundant shelly fauna that was listed by Wills and others (1925, p. 72-73) and Wills and Laurie (1938, p. 179-180). Ziegler checked the identifications of this material, which is preserved in the Department of Geology, University of Birmingham, Birmingham, England; the critical species are *Eocoelia curtisi* and *Costistricklandia lirata* var. *alpha*, which indicate a C$_5$ age. The highest beds of the Rubery Shales are not exposed, but by analogy with the nearby Walsall area they probably extend into C$_6$. The lower limit of the Rubery Shales is determined by the fact that the underlying Rubery Sandstone contains a C$_5$ fauna.

RUSHALL BEDS
(66)

These beds follow the Ludlow Bone Bed and are the lithologic equivalents of the Grey Downton Formation of the nearby Ludlow area (Squirrell and Tucker, 1960, p. 160–161).

SALOPIAN OF MEIFOD
(57)

The lowest Salopian beds of Meifod have yielded *Cyrtograptus murchisoni* (King, 1928, p. 691), but little is known of the rest of the Salopian sequence.

SALROCK FORMATION
(4)

See Killary Harbour Group.

SANDSTONE "SERIES" OF MARLOES
(46)

The Sandstone "Series" has not yielded a diagnostic fauna and has been tentatively correlated with the Ludlow (Cantrill and others, 1916, p. 52), presumably because of vague lithologic similarities with the Ludlow rocks of the Welsh Borderland. Since the underlying Coralliferous "Series" is largely of latest Llandovery age, it seems probable that the base of the Sandstone "Series" is of Wenlock age. There is a conformable passage with rocks of lower Old Red Sandstone type at Marloes, with some alternation of the two rock types at the junction. The top of the Sandstone "Series" has never been dated, but it may predate the end of the Ludlow (Walmsley, 1962).

SAUGH HILL GRITS
(18)

Cocks and Toghill (1973) have collected a single specimen of *Monograptus triangulatus* from these beds, indicating a general *M. gregarius* zone age.

SCART GRITS
(19)

These unfossiliferous grits have been correlated with the Saugh Hill Sandstone because of lithologic similarities (Lapworth, 1882, p. 643). They both overlie beds with a *M. cyphus* zone fauna.

SCOUT HILL FLAGS
(29)

Shaw (1971b, p. 608) has matched the ostracode fauna of these beds with the Downton of the Welsh Borderland. The critical element is the incoming of *Frostiella groenvalliana* at the base of the Scout Hill Flags and at the base of the Downton at Ludlow.

SEDGLEY LIMESTONE
(64)

King and Lewis (1912) have described this unit, often called the Aymestry Limestone, and Holland and others (1963, p. 150) concluded that both Bringewoodian and Leintwardinian shelly elements are present. Conodonts have been collected from this unit (Rhodes, 1953) and have been interpreted as a *siluricus* zonal assemblage (Fahraeus, 1969, p. 26-27).

SHALES, GRITS, AND CONGLOMERATES OF SLIEVE
BLOOM, KNOCKSHIGOWNA, BLESSINGTON, AND
CHAIR OF KILDARE
(15)

Harper (1948, p. 59) summarized the early literature on these poorly fossiliferous beds and concluded, from the lithology and the report of monograptids from Knockshigowna Hill, that the beds are of Silurian age. Recently, Rickards (*in* Cocks and others, 1971, p. 122) has identified *Monograptus riccartonensis* from Knockshigowna Hill, showing that some of the beds are of lower Wenlock age. Wright (1967, p. 42) pointed out that the unfossiliferous green grits that overlie the Ashgillian reef limestones at the Chair of Kildare must be of Silurian age. The lower Palaeozoic rocks of the Blessington area of Counties Kildare and Wicklow have been described by Brück (1970), who equated them with lithologically similar beds occurring above the Ashgillian at the Chair of Kildare. He recognized five formations which are, in ascending order, the Pollaphuca, Slate Quarries, Glen Ding, Tipperkevin, and Carrighill Formations. All these units are graywackes with the exception of the Slate Quarries Formation. Shelly fossils occur in the two top formations but are not diagnostic as to precise age (Brück, 1971). Acritarchs from the Slate Quarries Formation suggest an age of Middle Ordovician to Lower Silurian (Brück, 1971).

SHALES ON ANGLESEY
(32)

Greenly (1919, p. 481) gave a long list of graptolites identified by Elles from the Parys Mountain infold. Most of the specimens were collected from the spoil bank of a mine, and four zones are represented: *Monograptus cyphus, M. gregarius, M. convolutus,* and *M. sedgwicki.* Elles (*in* Greenly, 1919) believed that the overlying zone, *M. turriculatus,* might also be present because a fauna from another locality, this one collected in place, contained *M.* cf. *halli, M. jaculum, M.* cf. *nudus,* and *M. variabilis;* these identifications are uncertain because the material is badly preserved.

The relation of the Llandovery shales to the Ordovician rocks of the region is obscured by an igneous intrusion (Hawkins, 1966, p. 10, Fig. 2C). The graptolite zones at the top of the Ordovician and the zones at the bottom of the Llandovery are missing on Anglesey. Elles (*in* Greenly, 1919, p. 414) pointed out that this situation also persists at Conway on the mainland. At Conway, trilobite-bearing strata as well as barren strata intervene, whereas on Anglesey this position is occupied by the barren Parys Green Shales. Greenly (1919, p. 412) was inclined to correlate the Parys Green Shales with the "Barren Mudstones" of Moffat, which have in fact yielded uppermost Ordovician graptolites from a few shale seams. The Parys Green Shales could be of Llandovery age as well. In any case, there is no support in the literature for a stratigraphic break between Ordovician and Silurian on Anglesey.

Greenly (1919, p. 482) pointed out that "there is reason to believe that the succeeding members of the Silurian Series were largely composed of grit." These grits are known only as boulders in the Ligwy Bay Conglomerate of Carboniferous age. Greenly (p. 602) described these boulders as "hard siliceous grits, often quite coarse, highly pyritised, and manifestly from the metasomatic tract of Parys Mountain." No such rocks are present there now, and Greenly reasoned that they may be the remains of higher beds of the Silurian sequence. Fossils have not been found in these boulders, and they may be equivalent to the Upper Silurian grits of Denbighshire.

SHALE WITH SANDSTONE AND LIMESTONE OF THE BATSFORD OR LOWER LEAMINGTON BOREHOLE
(73)

Strahan (1913, p. 90-91) described these rocks, which were pierced to a depth of 49 meters below the angular unconformity with the Upper Coal Measures. Ziegler has re-examined the brachiopod fauna collected from this borehole (Ziegler and others, 1968b, p. 765) and discovered specimens of *Eocoelia sulcata* from many of the horizons. This, together with the fact that the lithologies represented are similar to the Rubery Sandstone and Rubery Shale, suggests a correlation of C_6 or, at the highest, low Wenlock for these beds.

SHANKEND BEDS
(26)

See Riccarton Group.

SHEERBATE FLAGS
(29)

See Coniston Grits.

SILTY SHALE OF THE BRABOURNE BOREHOLE
(76)

The Brabourne borehole penetrated "?Devonian or older" beds to a depth of 27 meters beneath the unconformity with the Triassic (Lamplugh and Kitchen, 1911, p. 54). The only fossil they reported was a ?Rhynchonellid. Cocks and others (1971, p. 125) reported that T. R. Lister has identified probable upper Llandovery chitinozoa from these beds.

SILURIAN OF CARDIFF
(70)

Sollas (1879, p. 479, 488) described these beds and reported lists of shelly fossils. He was inclined to correlate the beds with the Wenlock and Ludlow, although several of the fossils in his lists are restricted to the Llandovery, such as *Pentamerus oblongus* and *Stricklandia lirata*. Recently, D. A. Bassett (1969) has cast doubt on these identifications and has recorded *Monograptus flemingii* from the lowest beds seen. This would indicate a middle or upper Wenlock age, although the base of the sequence is unknown.

SKELGILL BEDS
(28, 29, 30)

See Stockdale Shales.

SKOMER VOLCANIC GROUP
(46)

Eocoelia hemisphaerica has been collected (Ziegler and others, 1969, p. 410) from the highest horizons of this unit and would indicate a C_1-C_2 age. Lower in the group, leperditicope ostracodes of Late Ordovician or Silurian type have been found recently (Ziegler and others, 1969), which would rule out the possibility of these volcanic rocks being Arenigian in age, a view favored by earlier workers. The lowest volcanic rocks seen are faulted against Llandeilo sediments.

SLATE QUARRIES FORMATION
(15)

See Shales, Grits, and Conglomerates of Slieve Bloom, Knockshigowna, Blessington, and Chair of Kildare.

SLATE QUARRY BEDS
(12)

See Little River Group.

SLATES AND GRITS FROM THE GALTY MOUNTAINS, SLIEVENAMAN, AND THE COMERAGH MOUNTAINS
(16)

Harper (1948, p. 58-59) has summarized the early work on these areas. He points out that *Monograptus riccartonensis* and a fossil resembling *Protovirgularia* have been reported from the Galty Mountains. These indicate an early Wenlock age. The lithologies of the areas are similar to Silurian lithologies elsewhere in central Ireland.

Cocks and others (1971, p. 122) recorded *M. flexilis* and *M. flemingii* high in the Galty Mountain succession, and these indicate a middle Wenlock age. They also reported (p. 123) an *Orthograptus* from the Silurian beds to the east of the Comeragh Mountains, which indicates a pre-late Llandovery age. In this area, the graptolitic Silurian is overlain by the Croughaun Beds.

SLATY ROCK OF CULFORD BOREHOLE
(83)

This rock was originally described as "?Silurian or older" by Whitaker (1906, p. 43). The drill penetrated about 6 meters beneath the unconformity with the Lower Greensand, and no fossils were found. A tentative lithologic correlation may be made with the Wenlock or upper Llandovery of the Stutton borehole.

SLOT BURN FORMATION
(20)

See Waterhead Group.

SMITHY BURN SILTSTONE
(21)

See Hagshaw Group.

SPECKLED GRIT
(69)

This unit is the lithologic equivalent of the Grey Downton Formation of the nearby Ludlow district (Walmsley, 1959, p. 510).

STOBS CASTLE BEDS
(26)

See Riccarton Group.

STOCKDALE SHALES
(28, 29, 30)

The Stockdale Shales have been carefully studied from the paleontologic point of view (Marr and Nicholson, 1888, p. 706; Rickards, 1970a), and a twofold subdivision into a lower unit, the Skelgill Beds, and an upper unit, the Browgill Beds, can be recognized throughout the Lake District and surrounding areas. At Skelgill itself, the Skelgill Beds have been subdivided into lower, middle, and upper units (Marr and Nicholson, 1888). The Lower Skelgill Beds in the Howgill Fells contain the zone of *Akidograptus acuminatus* and *Monograptus atavus* (Rickards, 1970a, p. 7, 8; Marr and Nicholson, 1888, p. 665, 679). Rickards (p. 6) has used the term Basal Beds for the strata that are intermediate between the Ashgillian and the graptolitic Llandovery and suggested that they may be approximately the age of the *Glyptograptus persculptus* zone. The *M. atavus* zone has been recognized in the Cross Fell inlier (Burgess and others, 1970, p. 170).

Marr and Nicholson (1888, p. 672) recognized three graptolite zones in the Middle Skelgill Beds, the *Monograptus fimbriatus* zone, the *M. argenteus* zone, and the *M. convolutus* zone. *M. cyphus* is listed as occurring in the first of these zones, so that the *M. fimbriatus* zone of Marr and Nicholson is the equivalent, at least in part, of the *M. cyphus* zone of the present terminology. Of these three graptolite zones, the second and third and possibly the first have been identified by King and Wilcockson (1934, p. 18-19) from the Skelgill Beds of the Austwick area. Recent work in the Cross Fell area (Burgess and others, 1970, p. 171-174) has established the *M. cyphus*, *M. triangulatus*, *D. magnus*, and *M. convolutus* zones. Rickards (1970a, p. 3) recorded these zones as well as the *M. acinaces* zone from the Howgill Fells.

Marr and Nicholson (1888, p. 674) distinguished two graptolite horizons in the Upper Skelgill Beds, the *M. clingani* band, and the zone of *M. spinigerus* (=*M. sedgwicki*). They gave a long list for the *M. clingani* band, which indicates the upper part of the zone of *M. convolutus*. In addition, some sections in the Howgill Fells, notably Spengill (Marr and Nicholson, p. 701, 708), yielded *Rastrites maximus* faunas, indicating the base of the zone of *Monograptus turriculatus*. In the Austwick area, King and Wilcockson (1934, p. 19) provisionally identified the zone of *M. sedgwicki* on the basis of the occurrence of *Climacograptus scalaris* and *Monograptus tenuis*.

The Lower Browgill Beds are characterized by the zones of *M. turriculatus* and *M. crispus*, and these zones have been identified in the Lake District (Marr and Nicholson, 1888, p. 678) and in the Cross Fell inlier (Shotton, 1935, p. 661). The Upper Browgill Beds have not yielded fossils in the Lake District, but the *M. griestoniensis* and *M. crenulata* zones have been collected in the Cross Fell area (Burgess and others, 1970, p. 175-176).

The main Lake District outcrop has been thoroughly revised by Hutt (1973).

STONEHAVEN BEDS
(24)

Campbell (1913, p. 930) divided the Stonehaven Beds into seven unnamed units; only the sixth unit, some 435 meters above the base of the sequence, proved fossiliferous. Westoll (1951, p. 6) has reviewed the evidence for the dating of this fauna, which consists of vertebrates and arthropods, and concludes that it indicates an early Downton (=Pridoli) age. The Stonehaven Beds rest with marked angular unconformity on Arenig volcanic and sedimentary rocks (Anderson, 1946, p. 486), so their lower limit could be well down in the Silurian as suggested by Lamont (1952, p. 30).

STRAITON GRITS
(18)

This unit has not yielded a diagnostic fauna but is presumed to be of Wenlock age because of its stratigraphic position.

STUDFOLD SANDSTONE
(30)

This unit is unfossiliferous (King and Wilcockson, 1934, p. 24) and is the probable equivalent of the Coniston Grits of the Lake District.

TALERDDIG GROUP
(38)

See Tarannon "Series."

TARANNON "SERIES"
(38)

Wood (1906, p. 655-664) first worked out the sequence of late Llandovery graptolite zones in this series, and in fact, the Tarannon area remains the one area in the British Isles where the sequence of zones is complete from middle Llandovery to Wenlock. Wood recognized four stratigraphic units in the Tarannon "Series": the Brynmair, the Gelli, the Talerddig, and the Dolgau Groups. These units are characterized by the zones of *M. turriculatus*, *M. crispus*, *M. griestoniensis*, and *M. crenulata*, respectively. The faunas representing these zones are large and constitute the definition of the zones. (The spelling of this name was changed recently to Trannon, but Tarannon is in the geologic literature.)

TEIRAN BEDS
(31)

Woods and Crosfield (1925, p. 176) collected a graptolite assemblage from these beds, which they regarded as transitional between the zones of *Monograptus nilssoni* and *M. scanicus*, and Warren (1967, written commun.) placed this assemblage in the upper part of the *M. nilssoni* zone.

TEMESIDE BEDS OF BIRMINGHAM
(64)

As the name would imply, these beds have been correlated with the Temeside Beds of the nearby Ludlow area, apparently because of general lithologic and faunal similarity (Ball, 1951, p. 232-233).

TEMESIDE BEDS OF LUDLOW
(63)

See Grey Downton Formation.

TEMESIDE SHALES OF BUCKNELL
(55)

A recent revision of this area is lacking, and all that can be said is that Stamp (1919, p. 234-236) intended a lithologic correlation with the Temeside Beds of the Ludlow area (see Grey Downton Formation).

TEMESIDE SHALES OF MALVERN
(65)

See Grey Downton Formation.

THIN GRITS OF LLANGRANOG AND ABERYSTWYTH
(43)

Hendriks (1926, p. 131, 134-135) collected several faunas from these grits, all of them indicative of the *Monograptus gregarius* zone.

THORNBURY BEDS
(71)

These beds succeed the Ludlow "Series" in the northern part of the Tortworth inlier and have yielded the early Ditton fish, *Traquairaspis* (Kellaway and Welch, 1948, p. 15). Presumably, therefore, they span the Downton interval.

TICKWOOD BEDS
(62)

See Wenlock Shales.

TILESTONES
(53)

These rocks were described by Straw (1929) and are the equivalents of the Long Quarry Beds, which occur along strike to the southwest (Potter and Price, 1965, p. 383). It should be noted that Straw used the term Tilestones in a more restricted sense than earlier workers, for example, Potter and Price (1965).

TIPPERKEVIN FORMATION
(15)

See Shales, Grits, and Conglomerates of Slieve Bloom, Knockshigowna, Blessington, and Chair of Kildare.

TOORMORE GROUP, INCLUDING TOORMORE CONGLOMERATE FORMATION, TOORMORE SANDSTONE FORMATION, AND TOORMORE BANDED FORMATION
(6)

Phillips and others (1970) recorded three formations from this group. The two basal formations, the Toormore Conglomerate Formation and the Toormore Sandstone Formation, occur only on Clare Island and are identical lithologically with the Cregganbaun Group of the adjacent Croagh Patrick synclinorium. The third formation, the Toormore Banded Formation, is known from both Clare Island and the Louisburgh area and contains five distinctive acid tuffs that also occur in the Oughty Group of the Croagh Patrick "Series." The Croagh Patrick "Series," in turn, has been correlated on lithologic grounds with the Wenlock rocks of northwest Galway. On Clare Island the Toormore Group rests unconformably on Dalradian metamorphic rocks, and in the Louisburgh area it is overlain conformably by the Knocknaveen Group.

TORTWORTH BEDS
(71)

Only the base of the Tortworth Beds is well exposed, and this contains what Curtis (1955b, p. 5-6) called the *Palaeocyclus* Band because of the great abundance of that coral. This would suggest a C_6 correlation for at least this band, and the presence of *Eocoelia sulcata* and *Costistricklandia lirata alpha* (Ziegler and others, 1968b, p. 763) strengthens this date. The contact with the overlying Wenlock

Series is not well defined because the two units are distinguished largely on lithologic grounds.

TRALORG FORMATION
(18)

Cocks and Toghill (1973) have revised the stratigraphy of the Girvan area and have collected a zonal assemblage representing the upper *M. cyphus* zone, including *M. cyphus* itself.

TRESGLEN BEDS
(48, 49, 50, 51, 52)

The only account of these beds is that of Price (in Lawson, 1966, p. 568) who described them as ". . . graptolitic shales of the zone of *Monograptus nilssoni*." Potter and Price (1965, p. 398) included these beds on their correlation chart; they made them the equivalent of the Lower Elton Beds of the Ludlow district.

TREWERN BROOK MUDSTONE FORMATION
(59)

The Wenlock graptolite zonation of the Long Mountain area has been studied in detail by Elles (1900, p. 386-397) in her classic paper and also by Das Gupta (1932, p. 326-333). Both authors presented long faunal lists to show that the three upper Wenlock graptolite zones are generally demonstrable in the various sections, that is, the zones of *Cyrtograptus linnarssoni, C. rigidus* (=*C. ellesae*), and *C. lundgreni.* The formation rests conformably on the Buttington Mudstone Formation, and recently Cocks and Rickards (1969, p. 226) reported the *C. centrifugus, C. murchisoni,* and *M. riccartonensis* zones from the basal beds. A brief description of this formation is given by Palmer (1970a, p. 343), who included a lenticular unit of shelly calcareous mudstone, the Glyn Member, within the formation. The correlation of this member with the graptolite zonal scheme was not established, however, in this introductory paper.

TRIANGULATUS BEDS
(39)

See Clywedog "Stage."

TRICHRÛG BEDS
(48, 49)

The Trichrûg Beds have not yielded a diagnostic fauna, but they are the lateral equivalents of the Lower Cwm Clŷd Beds which Potter and Price (1965, p. 399) correlated with the Upper Bringewood Beds of the type area on the basis of similarities of the shelly faunas.

TUFFS OF THE EASTERN MENDIPS
(72)

Reynolds (1907, p. 226-227) reported a shelly fauna from these beds, which was identified by Reed. Reed was of the opinion that the fauna was probably of late Llandovery age; his listing of *"Spirifer" crispus* definitely indicates a post-C$_1$ age. However, the absence of the common upper Llandovery brachiopods, a fact which Reed noted, suggests that the tuffs are probably younger, perhaps of Wenlock age. The base of the sequence is unknown, but the tuffs are succeeded, without angular break, by lava flows of pyroxene andesite (Reynolds, p. 234).

TURNER'S HILL BEDS
(64)

See Downton Castle Sandstone Group.

TWYMYN GROUP
(38)

Wood (1906, Table I) recognized the zones of *Cephalograptus cometa* and *M. sedgwicki* in this group. The zone of *Cephalograptus cometa* is usually taken as a subzone at the top of the zone of *Monograptus convolutus*.

TY-DRAW SLATES
(35)

Wills and Smith (1922, p. 198) reported the zones of *Monograptus crenulatus* (=*Monoclimacis crenulata*) and *M. griestoniensis* from the type locality of the Ty-draw Slates on the southern limb of the Llangollen syncline. Faunal lists for these zones are small and consist of *M. vomerinus* and *M. vomerinus* var. *crenulatus* for the upper zone, and *M. griestoniensis*, *M. marri?*, and *M. priodon* for the lower zone. On the north side of the Llangollen syncline, the presumed equivalents of the Ty-draw Slates have yielded larger faunas representing the zones of *M. turriculatus*, *M. crispus*, *M. griestoniensis*, and *M. crenulata* (Wills and Smith, p. 194-196).

UNIT M OF WALSALL, BIRMINGHAM
(64)

This unit was described by Butler (1937, p. 243, 246) as a gray mudstone and shale unit that rests above probable upper Llandovery purple and green shales, and beneath the Barr Limestone in the Walsall borehole. Unit M has yielded *Cyrtograptus murchisoni* and *Monograptus priodon* and so is assigned to the basal Wenlock zone of *Cyrtograptus murchisoni*.

"UPPER BIRKHILL" OF ABERGWESYN AND PUMPSAINT
(42)

Davies (1933, p. 184–187) has reported a large fauna representing the zone of *Monograptus sedgwicki* from these rocks.

UPPER BIRKHILL SHALES
(25)

See Birkhill Shales.

UPPER BLAISDON BEDS
(68)

These beds and the underlying Lower Blaisdon Beds have been compared with the Dayia Beds (Lower Leintwardine Beds) of Shropshire (Lawson, 1955, p. 104) on general lithologic and faunal grounds.

UPPER BODENHAM BEDS
(66)

These beds contain *Monograptus leintwardinensis* (Squirrell and Tucker, 1960, p. 148), and they have been correlated with the Upper Leintwardine Beds of Ludlow (Holland and others, 1959, p. 148). Squirrell and Tucker compared the Upper Bodenham Beds with the Lower Longhope Beds of May Hill and the Lower Llangibby Beds of Usk.

UPPER BRINGEWOOD BEDS
(62, 63)

These beds constitute mainly the Aymestry Limestone of former workers, and they were given their present name by Holland and others (1959). *Monograptus leintwardinensis* var. *incipiens* has been recorded (Holland and others, 1963, p. 113), suggesting a date immediately prior to the *M. leintwardinensis* zone. Earlier workers in the area (Wood, 1900, p. 428; Elles and Slater, 1906, p. 197; Alexander, 1936, p. 104) thought that the zone of *M. leintwardinensis* included the Aymestry Limestone and beds of earlier age; this conception may have been based on the fact that the type specimen of *M. leintwardinensis* came from Church Hill Quarry,

Leintwardine in beds beneath apparent Aymestry Limestone, but Whitaker (1962, Pl. XIV) has shown that these beds are Lower Leintwardine Beds and that the limestone is redeposited Aymestry Limestone boulders. A *siluricus* conodont zonal assemblage has been collected from the Upper Bringewood Beds (Rhodes, 1953; Rhodes and Newall, 1963; Fahraeus, 1969, p. 26).

UPPER BROWGILL BEDS
(29)

See Stockdale Shales.

UPPER CABAN CONGLOMERATE
(41)

See Caban Group.

UPPER CAMREGAN GRITS
(18)

Cocks and Toghill (1973) have collected a *M. turriculatus* zonal assemblage from these beds.

UPPER COLDWELL BEDS
(29)

See Upper Coniston Flags.

UPPER CONISTON FLAGS
(29)

The Lower, Middle, and Upper Coldwell Beds constitute the Upper Coniston Flags (Furness and others, 1967, p. 135). Watney and Welch (1910, p. 473) stated that the Lower Coldwell Beds, and also the top Lower Coniston Flags (Brathay Flags), contained the zone of *Cyrtograptus lundgreni*.

The Middle Coldwell Beds contain the zone of *Phacops obtusicaudatus*, which Watney and Welch (1910) regarded as equivalent to the zone of *Monograptus vulgaris*. Rickards (1965, p. 550) has renamed this trilobite *Delops nobilis marri*, and he has collected a *M. nilssoni-M. scanicus* zonal assemblage from the same beds. His fossils came from the upper part of the Middle Coldwell Beds (Rickards, 1967, written commun.). Blackie (1928, p. 97) reported a *C. lundgreni* assemblage from the Middle Coldwell Beds, but P. G. Llewellyn (1967, written commun.) thought that Blackie was mistaken and that the fossils came from the Lower Coniston

Flags. Rickards (1969, p. 69) has recently reported a *P. ludensis* zone assemblage from the lower part of the Middle Coldwell Beds. Watney and Welch (1910, p. 473) and Blackie (1933, p. 93) assigned the Upper Coldwell Beds to the zone of *Monograptus nilssoni*, and they also believed the overlying Coniston Grits extended down into the same zone. This zonal correlation has recently been confirmed by Furness and others (1967, p. 139–140).

UPPER CONISTON GRITS
(29)

See Coniston Grits.

UPPER CWM CLŶD BEDS
(48, 49, 50, 51, 52)

Potter and Price (1965, p. 399) correlated these beds with the Leintwardinian because of the presence of *Neobeyrichia lauensis* and *Calymene neointermedia*. It should be pointed out that this fauna came from the Llandeilo section where Potter and Price were unable to distinguish between the Upper Cwm Clŷd Beds and the Lower Roman Camp Beds. Neither of these species occurs in the Upper Cwm Clŷd Beds in areas where they can be clearly distinguished from the Lower Roman Camp Beds. Potter and Price regarded the Llandeilo section as a condensed one because the Upper Cwm Clŷd Beds and the Roman Camp Beds are much thinner than elsewhere, and they suggested that the lower part of the Upper Cwm Clŷd Beds are missing in this area. In the northern part of the Llandovery area, the Cwm Clŷd Beds are similar to the Wilsonia Shales of the Builth area (Potter and Price, p. 385).

UPPER ELTON BEDS
(62, 63)

This unit was named by Holland and others (1959) and is the upper part of the Lower Ludlow Shales of former workers. The only common graptolite is *Monograptus tumescens*. This graptolite also occurs in the Upper Elton Beds of the Wenlock area (Shergold and Shirley, 1968, p. 124).

UPPER FLAXLEY BEDS
(68)

Holland and others (1963, p. 149) assigned the Upper Flaxley Beds to the Bringewoodian Stage. Lawson (1955, p. 107) regarded the relation of the Upper Flaxley Beds to the Lower Flaxley Beds as probably unconformable and the relation to the succeeding Lower Blaisdon Beds, which contain a basal limestone conglomerate, as definitely unconformable.

UPPER FOREST BEDS
(69)

Holland and others (1963, Table II) have correlated these beds with the Lower Bringewood Beds of Ludlow, and Walmsley (1959, p. 508), who described them, compared them to the Upper Flaxley Beds of May Hill.

UPPER LEINTWARDINE BEDS OF TORTWORTH AND NEWNHAM
(71)

Mudstones and fine-grained sandstones in the Brookend borehole, at Tites Point and at Newnham have been correlated with the Upper Leintwardine Beds of Ludlow (Cave and White, 1971, p. 243-249; Curtis, 1972, p. 29).

UPPER LEINTWARDINE BEDS OF WENLOCK AND LUDLOW
(62, 63)

This unit yields the zonal graptolite *M. leintwardinensis* (Holland and others, 1959) as do the Lower Leintwardine Beds. These beds were usually included in the Dayia or Mocktree Shales by earlier workers. *M. leintwardinensis* also occurs in these beds in the Wenlock area in association with *M. leintwardinensis primus* (Shergold and Shirley, 1968, p. 128).

UPPER LLANBADOC BEDS
(69)

Walmsley (1959, p. 509) reported *Monograptus* cf. *leintwardinensis* var. *incipiens* from these beds and correlated them on general lithologic and faunal grounds with the Mocktree Shales (Lower Leintwardine Beds) of Ludlow. He also compared the Upper Llanbadoc Beds with the Upper Blaisdon Beds of May Hill because both have a basal limestone-conglomerate bed and are similar lithologically and faunally.

UPPER LLANDOVERY OF LLANDOVERY, LLANGADOCK, AND GARTH
(49, 50, 51, 52, 53)

The section in the central part of the southern Llandovery area, that is, around the River Ydw, was the first described and is selected here as the standard section of the upper Llandovery. O. T. Jones (1925b, p. 367-373) recognized six lithologic units, C_1 through C_6, and mapped the lowest unit as resting on various horizons ranging in age from B_3 to A_2. Ziegler (1966, p. 540-541) has collected *Eocoelia intermedia* and *Stricklandia lens ultima* from the C_4 horizon and *Eocoelia curtisi* from the C_5 horizon. Diagnostic graptolites have not yet been found in this sequence.

In the contiguous Sefin River section to the south, the sequence is greatly attenuated. Originally, Jones mapped all the upper Llandovery beds here as C_1, including a local development of black shale at the top of the more typical C_1 beds; it should be mentioned, however, that these "more typical" beds differ lithologically from the C_1 of the Ydw River section in that they are sandstone rather than mudstone. The sandstones of the Sefin River section are the type locality of *Stricklandia lens progressa* (Williams, 1951, p. 102), whereas the black shales have yielded *Eocoelia hemisphaerica* (Ziegler, 1966, p. 533). Moreover, 200 meters upstream from these localities, beds lithologically similar to the black shales have yielded *Monograptus sedgwicki* and *M. tenuis*, indicating a correlation with the zone of *M. sedgwicki* (O. T. Jones, 1925b, p. 370). Recently, we have collected these same graptolite species from the *Eocoelia* locality mentioned above. Also, in an appendix to Jones's 1949 paper, Jones and Williams described another set of strata between the black shale and the Wenlock. These beds were found to contain *Eospirifer radiatus* and *Clorinda globosa* and were thought to be equivalent with C_4 beds farther north, although their lithology, greenish sandy mudstone and calcareous sandstone, differs from the typical dark-gray sandstone of C_4.

Farther south, Williams (1953, p. 196-198) recognized upper Llandovery beds in the Llangadock area. He mapped these beds as resting on various horizons of the Llandeilo and Llanvirn and correlated them with the C_4-C_5 horizons of the type area. However, Ziegler (1966, p. 541) has collected *Eocoelia hemisphaerica* from the base of the sequence and *Eocoelia curtisi* and *Costricklandia lirata* var. *alpha* from near the top of the sequence, indicating that deposition began about C_1 or C_2 time and continued through C_5.

In the northern part of the Llandovery area, Jones (1949, p. 53-58) recognized three upper Llandovery units, which he called C_a, C_b, and C_c. He implied a lithologic correlation of the C_b sandstone bearing *Pentamerus oblongus* with C_4 sandstone of the southern area.

Farther north, at Garth, the sequence is slightly different. Andrew (1925, p. 397-399) mapped four units, C_a, C_b, C_c, and C_d, with unconformities at the base of C_a and apparently also at the base of C_d. Units C_a and C_b are conglomeratic sandstone and mudstone with abundant *Pentamerus* cf. *oblongus* and are similar in a general way to C_4 of the type area. The uppermost beds of C_d have yielded *Monograptus priodon* and *M. crenulatus*, suggesting a correlation with the zone of *M. crenulatus*. In this area, there is a gradation with the overlying Wenlock sequence, which contains a *Cyrtograptus murchisoni* fauna in its base.

Our section for the Llandovery of Garth and Builth is taken from the south side of the Silurian syncline, but on the north side, northwest of Builth Wells, there is exposed a narrow discontinuous belt of about 10 meters of sandstone referred to by Cocks and others (1971, p. 108-109) as the Trecoed Beds. These contain *Stricklandia lens ultima*, indicating a latest Llandovery age (Ziegler and others, 1968b, p. 669-670).

UPPER LLANDOVERY SHALES OF WALSALL, BIRMINGHAM
(64)

Butler (1937, p. 245-249) has provided a detailed account of these strata, the complete thickness of which was encountered in the Walsall borehole. The sequence is similar to that of Rubery Sandstone, also in the Birmingham district, except that the Rubery Sandstone, a thin basal sandstone unit, is not present at Walsall. Butler reported stricklandiids from a series of horizons extending from *Costistricklandia lirata* var. *alpha*, near the unconformity with the Cambrian quartzite, to *Costistricklandia lirata* var. *typica*. Butler was inclined to draw the upper limit of the upper Llandovery at the last occurrence of stricklandiids, but in this work the boundary is placed higher, at the upper limit of purple shale, because the purple coloration is typical of the uppermost Llandovery of Shropshire and other areas. *Cyrtograptus murchisoni* occurs above the uppermost purple shales in beds that Butler called "unit M," which indicates a basal Wenlock age for this unit.

The stricklandiids of Walsall indicate that the beds in which they occur range from C_5 to C_6, and the graptolites that Butler reported and Elles identified support this correlation. *Monograptus marri*, *M. discus*, and *M. priodon* (early var.) were reported from the lower part of the sequence, that is, the part that contains *Costistricklandia lirata* var. *alpha*; *M. vomerinus* var. *gracilis* and *M. vomerinus* or *M. vomerinus* var. *crenulatus* were reported from the beds with *Costistricklandia lirata* var. *typica*. The former list is composed of species that range through the later part of the late Llandovery, while the latter list is equivocal as to the latest Llandovery or earliest Wenlock graptolite zones. If the identification of *M. vomerinus crenulatus* is correct, then the *M. crenulatus* zone is indicated, whereas Butler quotes Elles as stating ". . . *M. vomerinus* var. *gracilis*, which is fairly common at the base of the Wenlock, is known also to occur in the *crenulatus* zone."

UPPER LLANGIBBY BEDS
(69)

This unit occurs just beneath the Downton Speckled Grit Beds and is thought to be of latest Ludlow age (Walmsley, 1959, p. 510).

UPPER LONGHOPE BEDS
(68)

Lawson (1955, p. 106) compared these beds with the Whitcliffe Flags of Ludlow, and he regarded them as resting unconformably on the Lower Longhope Beds because they contain a basal bed of phosphatized fragments.

UPPER LUDLOW FORMATION OF MALVERN
(65)

Phipps and Reeve (1967, p. 349-352) have recognized subdivisions of this unit, from bottom to top, as the Mocktree Shale Member, the Woodbury Shale Member, and the Whitcliffe Flags Member. These units thin considerably in the area south of Eastnor where they are represented by about 13 meters of conglomerate called the Clencher's Mill Member. The Mocktree Shale, Woodbury Shale, and Whitcliffe Flags Members are broadly equivalent to the Lower Leintwardine, Upper Leintwardine, and Whitcliffe Beds of the Ludlow district, respectively.

UPPER LUDLOW GROUP OF BIRMINGHAM
(64)

King and Lewis (1912) have provided a description of these beds, and Holland and others (1963, p. 150) have correlated them with the Whitcliffian of Ludlow.

"UPPER LUDLOW" OF CARDIFF
(70)

See Silurian of Cardiff.

UPPER NANTGLYN FLAG GROUP
(34)

The northwest Denbighshire area has been reinvestigated by P. T. Warren and his associates, of the Institute of Geological Sciences of Great Britain, and they regard the Upper Nantglyn Flag Group as the equivalent of the lower part of the *Monograptus nilssoni* zone.

UPPER OWENDUFF GROUP
(4)

This group has recently been restudied by Laird and McKerrow (1970), who subdivided it into the Lettergesh and Glencraff Formations. Rickards and Smyth (1968) have established the presence of the low Wenlock *C. murchisoni* and *M. riccartonensis* zones in the Lettergesh Formation. They reported *Monograptus flemingii* and *Monoclimacis flumendosae*, which would indicate that this formation extends into the middle Wenlock. The basal part of the Lettergesh Formation has been named the Gowlaun Member. In the western parts of this area, however, the lower beds of the Gowlaun Member may be older and equivalent in age to the Lower Owenduff Group (Piper, 1970, p. 509; Laird and McKerrow, 1970, p. 302).

UPPER PHOSPHATIZED PEBBLE BED
(67)

This unit was described by Lawson (1954, p. 233), who correlated it with the Ludlow Bone Bed.

UPPER RED DOWNTON FORMATION
(63)

See Red Downton Formation.

UPPER ROMAN CAMP BEDS
(50, 51, 52)

These beds have not yielded distinctive fossils, but are probably of Whitcliffian age because they overlie the Lower Roman Camp Beds of Leintwardinian age and underlie the Long Quarry Beds of supposed Downton age (Potter and Price, 1965, p. 399-400).

UPPER SAUGH HILL GRITS
(17)

These beds have not yielded diagnostic fossils but must be of middle Llandovery age because of their stratigraphic position (Cocks and Toghill, 1973).

UPPER SILTSTONES OF GORSLEY
(67)

Lawson (1954, p. 231-232) described these beds, which occur just beneath the Upper Phosphatized Pebble Bed of Downton age and Clifford's Mesne Sandstone. Holland and others (1963, p. 149) assigned these siltstones to the Whitcliffian.

UPPER SKELGILL BEDS
(29)

See Stockdale Shales.

UPPER SLEAVES OAK BEDS
(66)

This unit is the local representative of the Aymestry Limestone (Upper Bringewood Beds) and was correlated directly with the Upper Bringewood Beds of Ludlow by Holland and others (1963, Table II). A possible anomaly is the occurrence of *Monograptus leintwardinensis* in the Upper Sleaves Oak Beds (Squirrell and Tucker, 1960, p. 173). This unit is not present in the southeast portion of the Woolhope inlier because it was cut out by an erosional unconformity, and this is also the case in the contiguous areas to the southeast, Gorsley and May Hill (Squirrell and Tucker, p. 159). Squirrell and Tucker compared the Upper Sleaves Oak Beds with the Lower Llanbadoc Beds of the Usk inlier.

UPPER TRAP OF TORTWORTH
(71)

The Upper Trap is an extrusive lava that lies between the Damery Beds of C_5 age and the Tortworth Beds of C_6 age (Curtis, 1955b, p. 5). It pinches out to the north and west.

UPPER UNDERBARROW FLAGS
(29)

Shaw (1971a, p. 370) has matched the shelly fauna of these beds with the Whitcliffe Beds of the Ludlow district. Shaw (1971b, p. 607–609) regarded this unit as the facies equivalent of the lower part of the Kirkby Moor Flags and compared the ostracode fauna of both units with that of the Whitcliffe Beds of Ludlow.

"UPPER WENLOCK BEDS" OF DEVILSBIT MOUNTAIN DISTRICT
(2)

Cope (1959, p. 223) listed a composite graptolite fauna of *Cyrtograptus lundgreni, C. hamatus, Monograptus flemingii, M. dubius, M. vomerinus, Paraplectograptus eiseli,* and *Retiolites spinosus,* which indicates the *Cyrtograptus lundgreni* zone. The area is one of structural complexity, and the relation of these beds to the Cloncannon Beds, now known also to be of late Wenlock age, is unknown; Cope (p. 222) concluded that the gap is probably due to faulting. The base of the Upper Wenlock Beds is unknown.

"UPPER WENLOCK" OF LLANDEILO, LLANGADOCK, AND THE RIVER SEFIN
(48, 49, 50)

The only graptolite reported from these beds is *Monograptus* cf. *flemingi* (Williams, 1953, p. 198, 200), which, if the identification is correct, would indicate the *C. rigidus* or *C. lundgreni* zones of the Wenlock. Williams mapped these beds as resting unconformably on lower Wenlock and various horizons of the Ordovician. However, the evidence for an intra-Wenlock unconformity has been doubted (Cocks and others, 1971, p. 108). They contain a rich shelly fauna that is in accord with an upper Wenlock correlation.

UPPER WHITCLIFFE BEDS
(62, 63)

These beds were sometimes referred to as the *Chonetes* Flags by earlier workers (Holland and others, 1959) and directly underlie the Ludlow Bone Bed, so they are of latest Ludlow age. Recently, the *eosteinhornensis* conodont zone has been collected "from the upper part of the Whitcliffe Flags of Diddlebury, 6 miles north of Ludlow" (Collinson and Druce, 1966, p. 608).

UPPER WOOTTON BEDS
(66)

These beds were described by Squirrell and Tucker (1960, p. 145–146), who correlated them with the *Monograptus tumescens* zone (=*M. scanicus* zone) of present usage because of the abundance of that species in association with *M. chimaera* var. *salweyi*. They correlated the beds with the Lower Forest Beds of the Usk inlier, and Holland and others (1963, p. 148) correlated them with the Upper Elton Beds of the Ludlow area, mainly on the basis of the graptolite fauna.

UZMASTON BEDS
(44)

See Millin "Stage."

VENUSBANK FORMATION
(60)

Ziegler and others (1968b, p. 742) have introduced the term "Venusbank Formation" as a local name for strata that Whittard (1932, p. 874–878) called Pentamerus Beds. *Stricklandia lens* cf. *intermedia* or *progressa* has been collected from these beds (Ziegler and others, 1968b, p. 744), and Whittard (1932, p. 896) reported *Monograptus runcinatus* var. *pertinax*. These occurrences indicate a C_1 to C_2 correlation. Also, Ziegler and others (1968b) reported a graptolite identified by Rickards as *Climacograptus* aff. *rectangularis*. This graptolite indicates that beds younger than late Llandovery are present as well.

VIVOD GROUP
(35)

Monograptus leintwardinensis is the only fossil reported from these beds (Wills and Smith, 1922, p. 207; E.M.R. Wood, 1900, p. 446), and on this basis the Vivod Group is assigned to the zone of *M. leintwardinensis*.

WATERHEAD GROUP
(20)

The seven formations that compose this group are, from bottom to top, the Passage, Leaze, Birkenhead, Dippal Burn, Monument, Slot Burn, and Logan Burn Formations. The only diagnostic fossils are fish from the Dippal Burn Formation, and these are thought to be of late Wenlock or early to middle Ludlow age (Westoll, 1951, p. 6–7). Walton (1965, p. 199) suggested a correlation of this formation with the Fish Bed Formation of the Hagshaw Hills inlier.

WENLOCK LIMESTONE
(62, 63, 65, 66, 68, 69)

At the present time, the position of the Wenlock Limestone relative to the graptolitic scale is a matter of debate. Warren and others (1966) have shown that the

Pristiograptus ludensis zone is present in the Wenlock Limestone and the uppermost Wenlock Shales of the Ludlow district. Therefore, they regard the *P. ludensis* zone as the uppermost Wenlock zone and the *Pristiograptus nilssoni* zone as the basal Ludlow zone. Bassett and Shergold (1967) pointed out that the report of Das Gupta (1932, p. 351-352) indicated that *M. vulgaris* does occur in Lower Ludlow Shales of the adjacent area. However, *M. vulgaris* is known from the *P. nilssoni* zone elsewhere, and its record does not therefore necessarily affect the definition of the *P. ludensis* zone.

Unfortunately, there is no detailed information on the correlation of the Wenlock Limestone in the southern and eastern inliers of the Welsh Borderland. The graptolitic faunas of the underlying and overlying units have not been studied, so the only resort is to a lithologic correlation. Descriptions of the shelly fauna of the Wenlock Limestone have been given by Holland and others (1963, p. 105) for the Ludlow district, by Phipps and Reeve (1967, p. 344-346) for the Malvern district, by Squirrell and Tucker (1960, p. 172-178) for the Woolhope inlier, by Lawson (1955, p. 89-90) for the May Hill inlier, and by Walmsley (1959, p. 487-489) for the Usk inlier. Recently, a *sagitta* zone conodont fauna has been collected from the Usk inlier in beds that may be equivalent to or just below the basal Wenlock Limestone (Austin and Bassett, 1967, p. 276).

"WENLOCK LIMESTONE" OF CARDIFF
(70)

See Silurian of Cardiff.

"WENLOCK SERIES" OF FRESHWATER EAST
(47)

This term "Wenlock Series," as applied to the Freshwater East succession, refers to a small thickness of beds separated by unconformities from the underlying *Didymograptus bifidus* shales and the overlying "Ludlow Series" at Freshwater East (Dixon, 1921, p. 12). These beds are not present at Freshwater West, about 15 kilometers away, where the "Ludlow Series" rests directly on beds of Ordovician age. The "Wenlock Series" contains a slightly more diverse fauna than the "Ludlow Series," a situation that also prevails in areas to the north, such as Marloes, but it should be pointed out that the correlation of the Wenlock and Ludlow beds of the Marloes area is very uncertain.

WENLOCK SHALES OF KNIGHTON AND PRESTEIGNE
(54)

These beds have not been properly described, but Kirk (1951b, p. 72) mentioned that they were greatly attenuated in comparison with the Builth succession and that the basal Wenlock Shales contained *Cyrtograptus symmetricus* (Kirk, 1951a, p. 56).

WENLOCK SHALES OF LLANDOVERY, GARTH, AND BUILTH
(51, 52, 53)

The Builth area may be called the standard section for the Wenlock graptolite zonation, as it is the only area of several in the Welsh Borderland described by Elles (1900) in her classic paper that developed all six Wenlock zones. Furthermore, the sequence with both the upper Llandovery and lower Ludlow graptolite zones can be demonstrated within the area. The upper Llandovery to Wenlock sequence is seen in the Garth area and was described by Andrew (1925, p. 399), who obtained *Monograptus priodon* and *M. crenulatus* in the uppermost beds of his unit C_c of the upper Llandovery, while these species were joined by *Cyrtograptus murchisoni, Retiolites geinitzianus,* and *Monograptus capillaceous* in the overlying unit. He interpreted the first fauna as indicative of the zone of *M. crenulatus,* whereas the second indicates the zone of *Cyrtograptus murchisoni* or is a transition of the two. A similar faunal transition was described by O. T. Jones (1947a, p. 8-9) in the River Ithon area; here, however, the zonal indexes *Monograptus crenulatus* and *Cyrtograptus murchisoni* apparently do not occur together but always succeed each other in the expected order. The overlying "Lower Ludlow" Graptolitic Shales contain in their base the zone of *Monograptus ludensis,* which is now regarded as the top zone of the Wenlock.

The Wenlock Shales of the contiguous Llandovery area have never been thoroughly studied, although O. T. Jones (1925b, p. 375) reported the characteristic Wenlock form, *M. flemingii,* from these shales.

WENLOCK SHALES OF WELSHPOOL
(58)

Wade (1911, p. 441) reported a number of graptolite faunas from the Wenlock Shales, although none of these faunas consisted of more than three species. He assigned these faunas to four zones, *Monograptus riccartonensis, Cyrtograptus linnarssoni, C. rigidus* (=*C. ellesae*), and *C. lundgreni.* Of the two faunas assigned to the zone of *Monograptus riccartonensis,* one consisted entirely of the zonal index, and the other consisted of an anomalous assemblage, *Monograptus flemingi* var. *alpha, M. galaensis,* and *M. vomerinus* var. *alpha.* The fauna that Wade assigned to the *Cyrtograptus linnarssoni* zone consisted of the zone fossil in association with *Monograptus dubius* and *M. flemingi* var. *alpha.* The *Cyrtograptus rigidus* zone is represented at one locality by the zonal index questionably identified and at another locality by the joint occurrence of *Monograptus flemingi* var. *alpha* and *M. vomerinus* var. *alpha;* both, however, are long-ranging forms. Finally, Wade assigned several faunas to the *Cyrtograptus lundgreni* zone, apparently because of the occurrence of *Monograptus flemingi* var. *delta* and, at one locality, the occurrence of the zonal index.

Apparently then, many of the Wenlock horizons are present in the Welshpool area, even though it is difficult at present to ascertain, in most cases, the zone to which a fauna should be assigned. However, Wade's (1911, p. 437–438) mapping suggested some curious relations with the various Wenlock horizons. At Tyn-y-llwyn, he mapped the uppermost Wenlock zone, that of *Cyrtograptus lundgreni*, as resting directly on the Gaerfawr Beds of Ordovician age, while near Ty-brith he mapped the same zone as resting on the Cloddiau Group of Llandovery age. In this latter locality, he described "limestone conglomerates" from near the base of the sequence, lending support to his interpretation of a considerable unconformity at the base of the *C. lundgreni* zone. About 1 kilometer to the northwest and to the southeast of these localities, the earlier zones of the Wenlock (already described) occur. To explain the apparent relations, Wade proposed that an anticline was rising and eroding simultaneously. However, the evidence is too scanty now to fully accept such a hypothesis.

WENLOCK SHALES OF WENLOCK, LUDLOW, AND THE SOUTHERN PART OF THE WELSH BORDERLAND
(62, 63, 64, 65, 66, 68, 69)

Das Gupta (1932, p. 348–351) has made the most complete study of the graptolite zonation of the type Wenlock area; he recognized three stratigraphic units in the Wenlock Shales: the Buildwas Beds, the Coalbrookdale Beds, and the Tickwood Beds. We retain this threefold division of the Wenlock Shales while pointing out that the Geological Survey of Great Britain (Grieg and others, 1968, p. 148) restricts this term to the lower two units. In particular, Das Gupta collected from several localities near the base of the Wenlock Shales. He concluded that much of the sequence was missing, due to a possible unconformity with the underlying Purple Shales (Hughley Shales), as proposed by Whittard (1928, p. 752). Das Gupta thought that the three basal graptolite zones of the Wenlock were missing. His composite list from these beds is as follows: *Monograptus dubius, M. vomerinus, M. vomerinus* var. *basilicus, M. priodon, M. flemingi, M. flemingi* var. *primus, M. jaekeli?*, and *Gothograptus spinosus*. Unfortunately, he was unable to find identifiable cyrtograptids that are particularly useful in Wenlock zonation. Cocks and Rickards (1969, p. 225) have reinterpreted these lists and have concluded that they belong either to the *C. rigidus* or *C. linnarssoni* zones, or both. At the present time, Bassett, Cocks, Holland, Rickards, and Warren are engaged in a re-examination of the Wenlock of Wenlock Edge, and they have succeeded in collecting specimens of *Cyrtograptus centrifugus* from the lower Wenlock zone, thereby casting doubt on the existence of an unconformity at the base of the Wenlock.

The fauna mentioned above that was reported by Cocks and Rickards (1969) came from the lower part of the Coalbrookdale Beds; some came from the rather thin Buildwas Beds. Higher in the sequence, the *Cyrtograptus lundgreni* zone is well developed in a fairly great thickness of beds, which evidently includes the main part of the Coalbrookdale Beds and the overlying Tickwood Beds. It should be pointed out that Das Gupta equated the Coalbrookdale Beds with the *C. rigidus* zone in his correlation chart, but it is clear from his text and from Pocock and others (1938, p. 98) that *C. lundgreni* occurs well down in the Coalbrookdale Beds. At Ludlow, Holland and others (1969, p. 677) have recently reported the zones of *M. lundgreni* and *M. ludensis* from the uppermost Wenlock Shales.

There are two graptolite species recorded from the Birmingham area. Butler (1937, p. 246) reported *Monograptus priodon* from three horizons in the upper Wenlock Shales of the Walsall borehole, which indicates that these horizons are not later than the *Cyrtograptus rigidus* zone. At Dudley, Butler (1939, p. 55) collected *Monograptus flemingi* var. *delta* from just beneath the Dudley (Wenlock) Limestone. This variety occurs at a similar horizon in the type Wenlock area, according to Das Gupta (1932).

Virtually nothing is known of the graptolite zonation of the Wenlock Shale elsewhere in the Welsh Borderland. Descriptions of the Wenlock Shale in this region and lists of its shelly fauna have been given by Phipps and Reeve (1967, p. 344) for Malvern; by Squirrell and Tucker (1960, p. 143) for Woolhope; by Lawson (1955, p. 89) for May Hill; and by Walmsley (1959, p. 486) for Usk.

WERN QUARRY BEDS
(54)

These beds have not yielded diagnostic fossils (Holland, 1959b, p. 454) and Holland and others (1963, Table II) have assigned them to the Whitcliffian.

WHITCLIFFE BEDS OF TORTWORTH AND NEWNHAM
(71)

Mudstone, fine-grained sandstone, and limestone in the Brookend borehole, at Tites Point, and at Newnham, have been correlated with the Whitcliffe Beds of Ludlow (Cave and White, 1971, p. 243–249; Curtis, 1972, p. 29).

WHITCLIFFE FLAGS MEMBER OF MALVERN
(65)

See Upper Ludlow Formation of Malvern.

WILSONIA GRITS
(56)

Earp (1940, p. 3) has reported *Monograptus colonus, M. chimaera, M. tumescens, M. scanicus,* and *M. bohemicus* from the lower Wilsonia Grits of southwest Clun, and the first four of these species, as well as *M. uncinatus* var. *orbatus* and *M. uncinates* var. *micropoma,* were reported from the lower Wilsonia Grits of Kerry (Earp, 1938, p. 131). These identifications are all consistent with a *M.*

nilssoni-M. scanicus zone correlation. Also, in each of these areas, *M. leintwardinensis* var. *incipiens* occurs toward the top of the Wilsonia Grits (Earp, 1938). This same variety occurs also in what Earp called the Main Contorted Group, which is a local phenomenon of the southwest Clun area (Earp, 1940, p. 4). At Kerry, the Wilsonia Grits are overlain by the Monograptus Leintwardinensis Shales.

WILSONIA SHALES
(53)

The Wilsonia Shales probably range from the zone of *Monograptus nilssoni* to the zone of *M. leintwardinensis* because they contain (Straw, 1937, p. 450), in lower horizons, forms such as *M. bohemicus, M. chimaera* var. *salweyi*, and *M. colonus*, which suggest the *M. nilssoni* zone, whereas *M. leintwardinensis* occurs in the highest horizons.

WOOD BURN FORMATION
(18)

The top of the Wood Burn Formation contains *Monograptus sedgwicki* (Cocks and Toghill, 1973).

WOODBURY SHALE MEMBER OF MALVERN
(65)

See Upper Ludlow Formation of Malvern.

WOODLAND FORMATION
(19)

Boucot (Ziegler and others, 1966, p. 1033) has collected *Stricklandia lens* cf. *typica* from the Woodland Formation, suggesting a correlation with A_3-A_4. Cocks and Toghill (1973) have collected a graptolite fauna from these beds, which they interpret to represent the upper part of the *M. cyphus* zone.

WOOLHOPE LIMESTONE
(65, 66, 68)

The Woolhope Limestone directly overlies latest Llandovery beds with *Costistricklandia lirata* var. *typica* and *Eocoelia sulcata* at Woolhope and at May Hill, and in the Malvern district it overlies beds with *E. sulcata*. At May Hill, Ziegler has collected *E. sulcata* or *angelini* from sandstone beds within the thickness of the Woolhope Limestone (loc. no. 21, Ziegler and others, 1968b, p. 759). The exact correlation of the Woolhope Limestone is uncertain, and all that can be directly ascertained is that it must be of post-late Llandovery age. Lists of shelly fossils collected from the Woolhope Limestone are given by Squirrell and Tucker (1960, p. 172-178) for the Woolhope area, by Lawson (1955, p. 88-89) for the May Hill area, and by Phipps and Reeve (1967, p. 343-344) for the Malvern area.

WYCH BEDS
(65)

The Wych Beds overlie the Cowleigh Park Beds at Eastnor Park and Old Storridge Common, whereas to the east and southeast at Gullet Quarry and West Malvern, they rest directly on the Precambrian Malvernian (Ziegler, 1964). Farther south at Hollybush Quarry, they rest on the Cambrian Hollybush Sandstone (Jones and others, 1969). In all cases, *Eocoelia curtisi* has been collected from the basal beds, indicating a C_5 age (Ziegler and others, 1968b, p. 757). Also, a conodont fauna probably representing the *celloni* zone has been collected from the basal beds at Gullet Quarry (Brooks and Druce, 1965). At Hollybush Quarry a similar conodont fauna has been collected (Jones and others, 1969) which apparently occurs close to the boundary of the *celloni* and *amorphognathoides* zones. The top of the Wych Beds is not well exposed, but *Eocoelia sulcata*, together with *Palaeocyclus* sp., occurs near the top of the sequence at Alfrick, indicating a C_6 age. Immediately above this last mentioned locality, *Retiolites geinitzianus* var. *angustidens* has been found, which is known to range from the *Monograptus crispus* zone to the *Cyrtograptus centrifugus* zone.

YARTLETON BEDS
(68)

Lawson (1955, p. 88) has given a description of these beds, which lie above the early upper Llandovery Huntley Hill Beds and beneath the Woolhope Limestone. Ziegler and others (1968b, p. 760-761) reported *Costistricklandia lirata alpha* from the lower portion of the sequence and *C. lirata typica* and *Eocoelia sulcata* from the upper portion. These occurrences indicate a range of at least C_5 to C_6.

YELLOW DOWNTONIAN BEDS OF KNIGHTON AND LONG MOUNTAIN
(54, 59)

Holland (1959b, p. 462-464) described this unit at Knighton and regarded it as Downton on general lithologic and faunal grounds. This unit occurs above the Causemountain Formation at Long Mountain (Palmer, 1970a, p. 345) and has been matched with the succession of the Clun Forest area.

Y FRON-WERTH BEDS
(31)

Woods and Crosfield (1925, p. 174) collected a *Monograptus nilssoni* zone assemblage from this unit, which is the lowest unit exposed in the Clwŷdian Range.

YR ALLT GROUP
(58)

Wade (1911, p. 441) collected from several localities in this group, and most of the faunas are clearly *Monograptus nilssoni* zone assemblages. One collection consisted entirely of *M. vulgaris*, and Wade assigned this collection to the *M. vulgaris* (= *M. ludensis*) zone. Also, ?*M. leintwardinensis* was reported from one locality. Probably then, much of Ludlow time is represented by the Yr Allt Group.

YSTWYTH "STAGE"
(36, 37)

O. T. Jones (1909, p. 517) defined the Ystwyth "Stage" in the district around Plynlimon and Pont Erwyd and subdivided it into the Devil's Bridge Group, the Myherin Group, and the Rhuddnant Group. The faunas of these groups are not especially distinctive, however, and at best it can be said that *Monograptus turriculatus* occurs in the lowest and highest beds of the Ystwyth "Stage," suggesting that the highest beds in the area do not range above the zone of *M. crispus*. The lower limit is established by the fact that the underlying Pont Erwyd Stage extends into the zone of *M. sedgwicki*.

The Ystwyth "Stage" has also been recognized in the Machynlleth district, but faunas have not been reported from this area (Jones and Pugh, 1915, p. 365).

General Bibliography

A. M. ZIEGLER

This bibliography results from a relatively systematic search, over a fourteen-year period, for references pertaining to British and Irish Silurian stratigraphy, correlation, sedimentology, and paleoecology. The older papers have been included because many of them contain valuable observations on outcrops that no longer exist as well as interesting ideas that all too often have been forgotten. Most of the papers are stratigraphic accounts of lower Paleozoic areas, but general works, such as textbooks and the British Regional Geology series (most recent edition only), are included. Preliminary works, such as theses, abstracts, and notes in the Summary of Progress of the Geological Survey of Great Britain, are included if the information they contain has not been subsequently published. We have also included stratigraphic accounts of basically non-Silurian terrains if small patches of Silurian rocks are mentioned or if pebbles containing Silurian fossils from younger stratigraphic units are described. Field-trip accounts are included if they mention new finds, and papers containing claims of Silurian rocks that are dubious, such as those on Cornwall, are also listed. Papers of a strictly taxonomic nature have been left out unless they contain information of use in correlation. Papers on structure also were omitted unless they contain maps showing the distribution of Silurian rocks.

Each reference in the bibliography is followed by the name of the region within the British Isles to which the paper is devoted; this will aid the researcher who is unfamiliar with local place names. The regional categories and their relation to the columns on the correlation chart (Fig. 1) are as follows:

Ireland—columns 1-16;
Midland Valley (of Scotland)—columns 17-24;
Southern Uplands (of Scotland)—columns 25-27;
Northern England (Lake District and surrounding area)—columns 28-30;
North Wales—columns 31-35;
Central Wales—columns 36-43, 48-58;
Southwest Wales (mainly Pembrokeshire)—columns 44-47;
Northern Welsh Borderland—columns 59-64;
Southern Welsh Borderland—columns 65-72;
Southeast England (all boreholes)—columns 73-84;
Southwest England (these papers refer to dubious claims of Silurian strata that are not included on Fig. 1).

The following general categories are also used:

Scotland (general)—columns 17-27;
Wales and Welsh Borderland (general)—columns 31-72;
British Isles (general)—all columns;
Correlation (general)—information at the left-hand margin of Figure 1.

Adams, T. D., 1963, The geology of the Dinas Cwm-Rheidol hydroelectric tunnel: Geol. Mag., v. 100, p. 371-378. (Central Wales)

Aldridge, R. J., 1972, Llandovery conodonts from the Welsh Borderland: British Mus. (Nat. History) Bull., Geology, v. 22, p. 127-231. (Wales and Welsh Borderland—General)

Alexander, F.E.S., 1936, The Aymestry Limestone of the main outcrop: Geol. Soc. London Quart. Jour., v. 92, p. 103-115. (Northern Welsh Borderland)

Allen, J.R.L., and Tarlo, L. B., 1963, The Downtonian and Dittonian facies of the Welsh Borderland: Geol. Mag., v. 100, p. 129-155. (Northern Welsh Borderland)

Allender, R., Holland, C. H., Lawson, J. D., Walmsley, V. G., and Whitaker, J.H.McD., 1960, Summer field meeting at Ludlow: Geol. Assoc. London Proc., v. 71, p. 209-232. (Northern Welsh Borderland)

Anderson, J.G.C., 1946, The geology of the Highland Border: Stonehaven to Arran: Royal Soc. Edinburgh Trans., v. 61, p. 479-516. (Midland Valley)

——1958, Geology around the university towns: The Cardiff district: Geol. Assoc. London Guides No. 16: Norwich, Great Britain, Scientific Anglian, 12 p. (Southern Welsh Borderland)

——1960a, The Wenlock strata of south Mayo, Eire: Geol. Mag., v. 97, p. 265-275. (Ireland)

——1960b, Geology, in The Cardiff region: Cardiff, Wales Univ. Press, p. 22-44. (Southern Welsh Borderland)

——1963, The geology of the Rheidol hydroelectric project: Cardiganshire: South Wales Inst. Engineers Proc., v. 78, p. 35-47. (Central Wales)

Anderson, J.G.C., and Blundell, C.R.K., 1965, The sub-drift rock surface and buried valleys of the Cardiff district: Geol. Assoc. London Proc., v. 76, p. 367-378. (Southern Welsh Borderland)

Andrew, G., 1925, The Llandovery rocks of Garth, Breconshire: Geol. Soc. London Quart. Jour., v. 81, p. 389-406. (Central Wales)

Andrew, G., and Jones, O. T., 1925, The relations between the Llandovery rocks of Llandovery and those of Garth: Geol. Soc. London Quart. Jour., v. 81, p. 407-416. (Central Wales)

Anketell, J. M., 1963, The geology of the Llangranog district, southwest Cardiganshire [Ph.D. thesis]: Belfast, Queen's Univ. (Central Wales)

Anonymous, 1868, Penylan field meeting: Cardiff Nat. Soc. Ann. Rept., p. 38. (Southern Welsh Borderland)

Austin, J. E., 1925, Notes on the highest Silurian rocks of the Long Mountain: Geol. Assoc. London Proc., v. 36, p. 381-382. (Northern Welsh Borderland)

Austin, R. L., and Bassett, M. G., 1967, A sagitta Zone conodont fauna from the Wenlockian of the Usk inlier, Monmouthshire: Geol. Mag., v. 104, p. 274-283. (Southern Welsh Borderland)

Aveline, W. T., 1872, On the continuity and breaks between the various divisions of the Silurian strata of the Lake District: Geol. Mag., dec. 1, v. 9, p. 441-442. (Northern England)

——1876a, Absence of Llandovery rocks in the Lake District: Geol. Mag., dec. 2, v. 3, p. 282. (Northern England)

——1876b, The Silurian rocks of the Lake District (correspondence): Geol. Mag., dec. 2, v. 3, p. 376. (Northern England)

Aveline, W. T., and Hughes, T. McK., 1872, The geology of the country between Kendall, Sedbergh Bowness, and Tebag: Great Britain Geol. Survey Mem., sheet 98, N.E., old ser., 94 p. (Northern England)

Aveline, W. T., Hughes, T. McK., and Tiddeman, R. H., 1872, The geology of the neighbourhood of Kirkby Lonsdale and Kendall: Great Britain Geol. Survey Mem., sheet 98, S.E., old ser., 44 p. (Northern England)

Bailey, E. B., 1929, The Palaeozoic mountain systems of Europe and America: Rept. British

Assoc. Adv. Sci., Glasgow, 1928, Trans. Secs., p. 57-76. (British Isles—General)
——1930, New light on sedimentation and tectonics: Geol. Mag., v. 67, p. 77-92. (British Isles—General)
Bailey, E. B., and Holtedahl, O., 1938, Northwestern Europe Caledonides, in Andree, K., Brouwer, H. A., and Bucher, W. H., eds., Régionale Géologie der Erde: Leipzig, Akad. Verlag., v. 2, pt. 2, 76 p. (British Isles—General)
Bailey, R. J., 1964, A Ludlovian facies boundary in south central Wales: Geol. Jour., v. 4, p. 1-20. (Central Wales)
——1966, Scour ripples in the Ludlovian of south Radnorshire, Wales: Sedimentology, v. 7, p. 131-136. (Central Wales)
——1967, Paleocurrents and paleoslopes: A discussion: Jour. Sed. Petrology, v. 37, p. 1235-1255. (British Isles—General)
——1969, Ludlovian sedimentation in south central Wales, in Wood, A., ed., The Pre-Cambrian and lower Palaeozoic rocks of Wales: Cardiff, Wales Univ. Press, p. 283-304. (Central Wales)
Bailey, R. J., and Rees, A. I., 1973, A magnetic fabric study of late Whitcliffian siltstones from the Welsh Borderland: Geol. Jour., v. 8, p. 179-188. (Wales and Welsh Borderland—General)
Bailey, W. H., 1860, Notice of Upper Silurian fossils from Ballycar South, county of Clare, one mile and a half west of the village of Trough: Geol. Soc. Dublin Jour., v. 8, p. 109-110. (Ireland)
——1869, Notes on graptolites and allied fossils occurring in Ireland: Geol. Soc. London Quart. Jour., v. 25, p. 158-162. (Ireland)
Bailey, W. H., Jukes, J. B., and Kinahan, G. H., 1860, Explanations to accompany sheet 143 of the maps of the Geological Survey of Ireland, illustrating parts of the counties of Clare and Limerick: Geol. Survey Ireland Mem., 36 p. (Ireland)
Bailey, W. H., Jukes, J. B., and Wynne, A. B., 1860a, Explanations to accompany sheet 135 of the maps of the Geological Survey of Ireland, illustrating parts of Tipperary and of King's and Queen's Counties: Geol. Survey Ireland Mem., 32 p. (Ireland)
——1860b, Explanations to accompany sheet 145, of the maps of the Geological Survey of Ireland, illustrating part of the county of Tipperary: Geol. Survey Ireland Mem., 35 p. (Ireland)
——1861, Explanations to accompany sheet 134 of the maps of the Geological Survey of Ireland, illustrating parts of the counties of Clare, Tipperary, and Limerick: Geol. Survey Ireland Mem., 46 p. (Ireland)
Ball, H. W., 1951, The Silurian and Devonian rocks of Turner's Hill and Gornal, south Staffordshire: Geol. Assoc. London Proc., v. 62, p. 225-236. (Northern Welsh Borderland)
Ball, H. W., and Dineley, D. L., 1952, Notes on the Old Red Sandstone of the Clee Hills: Geol. Assoc. London Proc., v. 63, p. 207-214. (Northern Welsh Borderland)
Ball, H. W., and White, E. I., 1961, The Old Red Sandstone of Brown Clee Hill and the adjacent area. I. Stratigraphy: British Mus. (Nat. History) Bull., Geology, v. 5, p. 175-242. (Northern Welsh Borderland)
Banks, H. P., 1972, The stratigraphic occurrence of early land plants: Palaeontology, v. 15, p. 365-377. (British Isles—General)
Barrow, G., 1901, On the occurrence of Silurian rocks in Forfarshire and Kincardineshire: Geol. Soc. London Quart. Jour., v. 57, p. 328-345. (Midland Valley)
Bassett, D. A., 1955, The Silurian rocks of the Talerddig district, Montgomeryshire: Geol. Soc. London Quart. Jour., v. 111, p. 239-267. (Central Wales)
——1963, The Welsh Palaeozoic geosyncline: A review of recent work on stratigraphy and sedimentation, in Johnson, M.R.W., and Stewart, F. H., eds., The British Caledonides: Edinburgh and London, Oliver & Boyd, p. 35-69. (Wales and Welsh Borderland—General)

Bassett, D. A., 1969, Some of the major structures of early Palaeozoic age in Wales and the Welsh Borderland: A historical essay, *in* Wood, A., ed., The Pre-Cambrian and lower Palaeozoic rocks of Wales: Cardiff, Wales Univ. Press, p. 67-116. (Wales and Welsh Borderland—General)

Bassett, D. A., Whittington, H. B., and Williams, A., 1966, The stratigraphy of the Bala district, Merionethshire: Geol. Soc. London Quart. Jour., v. 122, p. 219-271. (Central Wales)

Bassett, M. G., 1969, The age of the oldest Silurian beds of the Rumney (Cardiff) inlier: Geol. Mag., v. 106, p. 90-92. (Southern Welsh Borderland)

——1971, Silurian rocks of the south Pembrokeshire coast, *in* Bassett, D. A., and Bassett, M. G., eds., Geological excursions in South Wales and the Forest of Dean: Cardiff, Wales Univ. Press, p. 206-221. (Southwest Wales)

Bassett, M. G., and Shergold, J. H., 1967, The position of the Wenlock/Ludlow boundary in the Silurian graptolite sequence: Geol. Mag., v. 104, p. 395-397. (Correlation—General)

Bassett, M. G., Cocks, L.R.M., Holland, C. H., Rickards, R. B., and Warren, P. T., 1974, The type Wenlock Series: Geol. Soc. London Jour. (in press). (Northern Welsh Borderland)

Bather, F. A., 1907, The discovery in west Cornwall of a Silurian crinoid characteristic of Bohemia: Royal Geol. Soc. Cornwall Trans., v. 13, p. 192. (Southwest England)

Beloussov, V., and Gsovsky, M. V., 1945, Geosynclines, their structure, history, and laws of development. The Caledonian geosyncline of Great Britain: Soc. Nat. Moscow Bull., Sec. Geol., new ser., v. 20, p. 130-164. (British Isles—General)

Bennett, A., 1942, The geology of Malvernia: Malvern, England, Malvern Naturalists' Field Club, unpaginated. (Southern Welsh Borderland)

Bennison, G. M., and Wright, A. E., 1969, The geological history of the British Isles: New York, St. Martins Press, 406 p. (British Isles—General)

Berry, W.B.N., and Boucot, A. J., 1970, Correlation of the North American Silurian rocks: Geol. Soc. America Spec. Paper 102, 289 p. (British Isles—General)

——1972a, Correlation of the South American Silurian rocks: Geol. Soc. America Spec. Paper 133, 59 p.

——1972b, Correlation of the southeast Asian and Near Eastern Silurian rocks: Geol. Soc. America Spec. Paper 137, 65 p.

——1973, Correlation of the African Silurian rocks: Geol. Soc. America Spec. Paper 147, 83 p.

Black, M., 1957, Sedimentation in relation to the Caledonian movements in Britain: Internat. Geol. Cong., 20th, Mexico [D.F.] 1956, Rept., pt. 5, p. 139-152. (British Isles—General)

Blackie, R. C., 1927a, The geology of the country between Llanelidan and Bryneglwys: Geol. Soc. London Quart. Jour., v. 83, p. 711-735. (North Wales)

——1927b, The geology of the Llanelidan district, North Wales: Liverpool Geol. Soc. Proc., v. 14, p. 306-318. (North Wales)

——1928, The geology of the southern end of the Clwydian Range: Liverpool Geol. Soc. Proc., v. 15, p. 21-55. (North Wales)

——1933, The Silurian rocks of the Kentmere district, Westmorland: Liverpool Geol. Soc. Proc., v. 16, p. 88-105. (Northern England)

Bloxam, T. W., 1971, Haverfordwest, Strumble Head, and Abereiddy Bay, *in* Bassett, D. A., and Bassett, M. G., eds., Geological excursions in South Wales and the Forest of Dean: Cardiff, Wales Univ. Press, p. 199-205. (Southwest Wales)

Blyth, F.G.H., 1952, Malvern tectonics—A contribution: Geol. Mag., v. 89, p. 185-194. (Southern Welsh Borderland)

Blyth, F.G.H., and Blackith, R. E., 1953, A temporary section in the Malverns: Geol. Mag., v. 90, p. 442-443. (Southern Welsh Borderland)

Bolton, J., 1869, Geological fragments—Collected principally from rambles among the rocks of Furness and Cartmel: London, Whittaker and Co., 264 p. (Northern England)

Boswell, P.G.H., 1926, A contribution to the geology of the eastern part of the Denbighshire Moors: Geol. Soc. London Quart. Jour., v. 82, p. 556-584. (North Wales)

——1927, The Salopian rocks and tectonics of the district south-west of Ruthin (Denbighshire): Geol. Soc. London Quart. Jour., v. 83, p. 689-710. (North Wales)

——1928, The cleavage-fan in the Silurian rocks of the Denbighshire Moors and Clwydian Range: Liverpool Geol. Soc. Proc., v. 15, p. 69-77. (North Wales)

——1930, The pre-Carboniferous history of the Vale of Clwŷd: Liverpool Geol. Soc. Proc., v. 15, p. 230-240. (North Wales)

——1931, The Ludlow rocks of the northern part of the Clwŷdian Range: Liverpool Geol. Soc. Proc., v. 15, p. 297-308. (North Wales)

——1932, On the occurrence and significance of an area of imbrication in the Ludlow rocks of the Denbighshire Moors: Liverpool Geol. Soc. Proc., v. 16, p. 18-32. (North Wales)

——1935a, The geology of north-western Denbighshire: Geol. Assoc. London Proc., v. 46, p. 152-186. (North Wales)

——1935b, Report of Easter Field Meeting, 1935, north-western Denbighshire: Geol. Assoc. London Proc., v. 46, p. 193-202. (North Wales)

——1937, The tectonic problems of an area of Salopian rocks in north-western Denbighshire: Geol. Soc. London Quart. Jour., v. 93, p. 284-321. (North Wales)

——1942, The Wenlock and Ludlow rocks of the district around Gwytherin, north-western Denbighshire: Liverpool Geol. Soc. Proc., v. 18, p. 86-100. (North Wales)

——1943a, The Salopian rocks and geological structure of the country around Eglwys-fach and Glan Conway, N.W. Denbighshire: Geol. Assoc. London Proc., v. 54, p. 93-112. (North Wales)

——1943b, A revision of the area around Llyn-Goronwy Llanrwst, Denbighshire: Liverpool Geol. Soc. Proc., v. 18, p. 144-148. (North Wales)

——1945, The occurrence of the zone of *Cyrtograptus rigidus* (Wenlock Series) in North Wales, with notes on the age of the Denbigh Grit Series: Liverpool Geol. Soc. Proc., v. 19, p. 72-96. (North Wales)

——1949, The Middle Silurian rocks of North Wales: London, Edward Arnold and Co., 448 p. (North Wales)

——1953, The alleged subaqueous sliding of large sheets of sediment in the Silurian rocks of North Wales: Geol. Jour., v. 1, p. 148-152. (North Wales)

——1956, The Middle Silurian rocks of North Wales: Some notes and corrections: Geol. Jour., v. 1, p. 326-327. (North Wales)

——1961, The case against a lower Palaeozoic geosyncline in Wales: Geol. Jour., v. 2, p. 612-624. (Wales and Welsh Borderland—General)

Boswell, P.G.H., and Double, I. S., 1934, The Ludlow rocks of the northern part of the Denbigh Moors between Abergele and Llanfair Talhaiarn: Liverpool Geol. Soc. Proc., v. 16, p. 156-172. (North Wales)

——1938, The Ludlow rocks and structure of the country in the neighborhood of Llanfair-Talhaiarn and Llansannan, Denbighshire: Liverpool Geol. Soc. Proc., v. 17, p. 277-311. (North Wales)

——1940, The geology of an area of Salopian rocks west of the Conway Valley, in the neighborhood of Llanrwst, Denbighshire: Geol. Assoc. London Proc., v. 51, p. 151-187. (North Wales)

Boulton, W. S., 1928, The geology of the northern part of the Lickey Hills, near Birmingham: Geol. Mag., v. 65, p. 255-266. (Northern Welsh Borderland)

Bowman, J. E., 1838, Notes on a small patch of Silurian rocks to the west of Abergele, on the north coast of Denbighshire: Geol. Soc. London Proc., v. 2, p. 666-667. (North Wales)

Bowman, J. E., 1841, Notice of Upper Silurian rocks in the Vale of Llangollen, North Wales, and of a contiguous eruption of trap and compact feldspar: Manchester Geol. Assoc. Trans., v. 1, p. 194-211. (North Wales)

Brodie, P. B., 1871, On the "passage-beds" in the neighborhood of Woolhope, Herefordshire, and on the discovery of a new species of *Eurypterus* and some land-plants in them: Geol. Soc. London Quart. Jour., v. 27, p. 256-263. (Southern Welsh Borderland)

Brooks, M., 1968, The geological results of gravity and magnetic surveys in the Malvern Hills and adjacent districts: Geol. Jour., v. 6, p. 13-30. (Southern Welsh Borderland)

——1970, Pre-Llandovery tectonism and the Malvern structure: Geol. Assoc. London Proc., v. 81, p. 249-268. (Southern Welsh Borderland)

Brooks, M., and Druce, E. C., 1965, A Llandovery conglomeratic limestone in Gullet Quarry, Malvern Hills, and its conodont fauna: Geol. Mag., v. 102, p. 370-382. (Southern Welsh Borderland)

Brown, D. J., 1872a, On the Silurian rocks of the south of Scotland: Rept. British Assoc. Adv. Sci., Edinburgh, 1871, Trans. Secs., p. 93. (Southern Uplands)

——1972b, On the Upper Silurian rocks of the Pentland Hills and Lesmahgo: Rept. British Assoc., Adv. Sci., Edinburgh, 1871, Trans. Secs., p. 93. (Midland Valley)

——1874a, On the Silurian rocks of the south of Scotland. Pt. I—Moffat and Gala beds: Edinburgh Geol. Soc. Trans., v. 2, p. 227-237. (Scotland—General)

——1874b, On the Silurian rocks of the south of Scotland. Pt. II—Llandovery rocks: Edinburgh Geol. Soc. Trans., v. 2, p. 216-321. (Scotland—General)

——1874c, On the Silurian rocks of the south of Scotland. Pt. III—Wenlock and Ludlow rocks: Edinburgh Geol. Soc. Trans., v. 2, p. 377-383. (Scotland—General)

Brown, D. J., and Henderson, J., 1867, On the Silurian rocks of the Pentland Hills, with notes on the Brachiopoda by T. Davidson: Edinburgh Geol. Soc. Trans., v. 1, p. 23-33. (Midland Valley)

Brück, P. M., 1970, Stratigraphy, petrology, and structure of the greywacke formations of the Blessington area: Geol. Survey Ireland Bull., v. 1, p. 31-45. (Ireland)

——1971, Fossil content and age of the greywacke formations west of the Leinster Granite in Counties Dublin, Kildare, and Wicklow, Ireland: Geol. Mag., v. 108, p. 303-310. (Ireland)

——1972, Stratigraphy and sedimentology of the lower Palaeozoic greywacke formations in Counties Kildare and West Wicklow: Royal Irish Acad. Proc., v. 72, sec. B, p. 25-53. (Ireland)

Buckman, S. S., 1904, Some Ludlovian brachiopods and a question about Silurian time: Geol. Assoc. London Proc., v. 18, p. 454-458. (Correlation—General)

Bullard, E. C., Gaskell, T. F., Harland, W. B., and Kerr-Grant, C., 1940, Seismic investigations of the Palaeozoic floor of east England: Royal Soc. London Philos. Trans., ser. A, v. 239, p. 29-94. (Southeast England)

Burgess, I. C., Rickards, R. B., and Strachan, I., 1970, The Silurian strata of the Cross Fell area: Great Britain Geol. Survey Bull., no. 32, p. 167-182. (Northern England)

Butcher, N. E., 1962, The tectonic structure of the Malvern Hills: Geol. Assoc. London Proc., v. 73, p. 103-125. (Southern Welsh Borderland)

Butler, A. J., 1937, On the Silurian and Cambrian rocks encountered in a deep boring at Walsall, south Staffordshire: Geol. Mag., v. 74, p. 241-257. (Northern Welsh Borderland)

——1939, The stratigraphy of the Wenlock Limestone of Dudley: Geol. Soc. London Quart. Jour., v. 95, p. 37-74. (Northern Welsh Borderland)

Butler, A. J., and Oakley, K. P., 1936, Report of "Coral Reef" meeting at Wenlock Edge, the Dudley District, and the Oxford District: The Dudley District: Geol. Assoc. London Proc., v. 47, p. 133-136. (Northern Welsh Borderland)

Callaway, C., 1879, The tripartite division of the Silurian and Cambrian formations: Geol.

Mag., dec. 2, v. 6, p. 142-143. (Correlation—General)

Campbell, R., 1911, Preliminary note on the geology of S.E. Kincardineshire: Geol. Mag., dec. 5, v. 8, p. 63-69. (Midland Valley)

——1912a, The Downtonian and Old Red Sandstone of Kincardineshire: Rept. British Assoc. Adv. Sci., Dundee, 1912, Trans. Secs., p. 461-462. (Midland Valley)

——1912b, Cambrian, Downtonian, and lower Old Red Sandstone rocks near Stonehaven: Geol. Assoc. London Proc., v. 23, p. 291-294. (Midland Valley)

——1912c, On the geology of south-eastern Kincardineshire: Geol. Assoc. London Proc., v. 23, p. 295-298. (Midland Valley)

——1913, The geology of southeastern Kincardineshire: Royal Soc. Edinburgh Trans., v. 48, p. 923-960. (Midland Valley)

——1914, The Pentland Hills: Geol. Assoc. London Proc., v. 25, p. 19-24. (Midland Valley)

——1927, The geology of the district around Edinburgh; the neighborhood of Edinburgh. IV. The Pentland Hills: Geol. Assoc. London Proc., v. 38, p. 427-431. (Midland Valley)

——1929, The composition of the conglomerates of the Downtonian and lower Old Red Sandstone of the Stonehaven district: Rept. British Assoc. Adv. Sci., Glasgow, 1928, Trans. Secs., p. 554-555. (Midland Valley)

Cantrill, T. C., 1902, Report on the examination of the sections at the reservoirs and along the course of the aqueduct of the Birmingham Corporation Waterworks, Rhayader to Frankley, 1901: Great Britain Geol. Survey Summary Progress for 1901, p. 60-64. (Central Wales)

——1907, Stratigraphical note: Geol. Mag., dec. 5, v. 4, p. 537-538. (Southwest Wales)

——1927, Midlands district: Great Britain Geol. Survey Summary Progress for 1926, p. 41-45. (Northern Welsh Borderland)

Cantrill, T. C., Dixon, E.E.L., Thomas, H. H., and Jones, O. T., 1916, Geology of the South Wales coalfield. Pt. XII. The country around Milford: Great Britain Geol. Survey Mem., sheet 227, new ser., 185 p. (South Wales)

Capewell, J. G., 1957, The stratigraphy, structure, and sedimentation of the Old Red Sandstone of the Cameragh Mountains and adjacent areas, County Waterford, Ireland: Geol. Soc. London Quart. Jour., v. 112, p. 393-412. (Ireland)

Carruthers, R. G., and Muff, H. B., 1909, The lower Palaeozoic rocks around Killary Harbour: Irish Naturalists' Jour., v. 11, p. 7-11. (Ireland)

Cave, R., 1955, The stratigraphy of the Welshpool area (Montgomeryshire) [Ph.D. thesis]: Cambridge, Cambridge Univ. (Central Wales)

——1965, The Nod Glas sediments of Caradoc age in North Wales: Geol. Jour., v. 4, p. 279-298. (North Wales)

Cave, R., and White, D. E., 1968, Brookend borehole: Great Britain Geol. Survey Summary Progress for 1967, p. 75-76. (Southern Welsh Borderland)

——1971, The exposures of Ludlow rocks and associated beds at Tites Point and near Newnham, Gloucestershire: Geol. Jour., v. 7, p. 239-254. (Southern Welsh Borderland)

Challinor, J., 1928, A shelly band in graptolitic shales: Geol. Mag., v. 65, p. 364-368. (Central Wales)

——1949, The origin of certain rock structures near Aberystwyth: Geol. Assoc. London Proc., v. 60, p. 48-53. (Central Wales)

——1969, A review of geological research in Cardiganshire, 1842-1967: Welsh Geol. Quart., v. 4, p. 3-40. (Central Wales)

Charlesworth, H.A.K., 1960, The lower Palaeozoic inlier of the Curlew Mountains anticline: Royal Irish Acad. Proc., v. 61, sec. B, p. 37-50. (Ireland)

Charlesworth, J. K., 1935, The geology of north-east Ireland: Geol. Assoc. London Proc., v. 46, p. 441-492. (Ireland)

Charlesworth, J. K., 1963, Historical geology of Ireland: Edinburgh and London, Oliver & Boyd, 565 p. (Ireland)

——1966, The geology of Ireland: Edinburgh and London, Oliver & Boyd, 276 p. (Ireland)

Cobbold, E. S., 1900, Geology, in Campbell-Hyslop, C. W., ed., Vol. 1, Church Stretton: Shrewsbury, L. Wilding, 115 p. (Northern Welsh Borderland)

——1904, Unconformities in the Church Stretton district: Geol. Assoc. London Proc., v. 18, p. 442-443. (Northern Welsh Borderland)

——1925, Unconformities in south Shropshire: Geol. Assoc. London Proc., v. 36, p. 364-367. (Northern Welsh Borderland)

Cocks, L.R.M., 1967a, Depth patterns in Silurian marine communities: Marine Geology, v. 5, p. 379-382. (British Isles—General)

——1967b, Llandovery stropheodontids from the Welsh Borderland: Palaeontology, v. 10, p. 245-265. (Wales and Welsh Borderland—General)

——1968, Some strophomenacean brachiopods from the British Lower Silurian: British Mus. (Nat. History) Bull., Geology, v. 15, p. 285-324. (Correlation—General)

——1970, Silurian brachiopods of the superfamily Plectambonitacea: British Mus. (Nat. History) Bull., Geology, v. 19, p. 141-203. (Correlation—General)

——1971a, Facies relationships in the European Lower Silurian, in Babin, C., ed., Colloque ordovicien-silurien: Bur. Recherches Géol. et Minières Mem. no. 73, p. 223-228. (British Isles—General)

——1971b, The Llandovery district, in Bassett, D. A., and Bassett, M. G., eds., Geological excursions in South Wales and the Forest of Dean: Cardiff, Wales Univ. Press, p. 155-161. (Central Wales)

Cocks, L.R.M., and Rickards, R. B., 1969, Five boreholes in Shropshire and the relationships of shelly and graptolitic facies in the Lower Silurian: Geol. Soc. London Quart. Jour., v. 124, p. 213-238. (Northern Welsh Borderland)

Cocks, L.R.M., and Toghill, P., 1973, The biostratigraphy of the Silurian rocks of the Girvan district, Scotland: Geol. Soc. London Jour., v. 127, p. 209-243. (Midland Valley)

Cocks, L.R.M., and Walton, G., 1968, A large temporary exposure in the Lower Silurian of Shropshire: Geol. Mag., v. 105, p. 390-397. (Northern Welsh Borderland)

Cocks, L.R.M., Toghill, P., and Ziegler, A. M., 1970, Stage names within the Llandovery Series: Geol. Mag., v. 107, p. 79-87. (Correlation—General)

Cocks, L.R.M., Holland, C. H., Rickards, R. B., and Strachan, I., 1971, A correlation of Silurian rocks in the British Isles: Geol. Soc. London Quart. Jour., v. 127, p. 103-136. (British Isles—General)

Cole, G.A.J., 1918, Gothlandian or Silurian. B. Ireland, in Evans, J. W., ed., Geology of the British Isles: The Hague, Martins Nijhoff, p. 103-134. (Ireland)

Cole, G.A.J., and Hallissy, T., 1924, Handbook of the geology of Ireland: London, Thomas Murby and Co., 82 p. (Ireland)

Collinson, C., and Druce, E. C., 1966, Upper Silurian conodonts from Welsh Borderlands [abs.]: Am. Assoc. Petroleum Geologists Bull., v. 50, p. 608. (Northern Welsh Borderland)

Cope, R. N., 1954, Cyrtograptids and retiolitids from County Tipperary: Geol. Mag., v. 91, p. 319-324. (Ireland)

——1959, The Silurian rocks of the Devilsbit Mountain district, County Tipperary: Royal Irish Acad. Proc., v. 60, sec. B, p. 217-242. (Ireland)

Craig, G. Y., and Walton, E. K., 1959, Sequence and structure in the Silurian rocks of Kirkcudbrightshire: Geol. Mag., v. 96, p. 209-220. (Southern Uplands)

——1962, Sedimentary structures and palaeocurrent directions from the Silurian rocks of Kirkcudbrightshire: Edinburgh Geol. Soc. Trans., v. 19, p. 100-119. (Southern Uplands)

Crewdson, G., 1915, New fossiliferous horizon in the Coniston Grits of Windermere: Geol. Mag., dec. 6, v. 2, p. 169-171. (Northern England)

Crosfield, M. C., and Johnston, M. S., 1914, A study of ballstone and the associated beds in the Wenlock Limestone of Shropshire: Geol. Assoc. London Proc., v. 25, p. 193-224. (Northern Welsh Borderland)

Cruise, R. J., and Bailey, W. H., 1885, Explanatory memoir to accompany sheet 58 of the maps of the Geological Survey of Ireland, illustrating parts of Armagh, Fermanagh, and Monaghan: Geol. Survey Ireland Mem., 32 p. (Ireland)

Cummins, W. A., 1957, The Denbigh Grits; Wenlock greywackes in Wales: Geol. Mag., v. 94, p. 435-451. (Wales and Welsh Borderland—General)

——1959a, The Nantglyn Flags; mid-Salopian basin facies in Wales: Geol. Jour., v. 2, p. 159-167. (Wales and Welsh Borderland—General)

——1959b, The Lower Ludlow Grits in Wales: Geol. Jour., v. 2, p. 168-179. (Wales and Welsh Borderland—General)

——1962, The greywacke problem: Geol. Jour., v. 3, p. 51-72. (British Isles—General)

——1969, Patterns of sedimentation in the Silurian rocks of Wales, in Wood, A., ed., The Pre-Cambrian and lower Palaeozoic rocks of Wales: Cardiff, Wales Univ. Press, p. 219-237. (Wales and Welsh Borderland—General)

Curtis, M.L.K., 1955a, A review of past research on the lower Palaeozoic rocks of the Tortworth and Eastern Mendip inliers: Bristol Naturalists' Soc. Proc., v. 29, p. 71-78. (Southern Welsh Borderland)

——1955b, Lower Palaeozoic, in Bristol and its adjoining countries: Bristol, J. W. Arrowsmith Ltd., p. 3-7. (Southern Welsh Borderland)

——1972, The Silurian rocks of the Tortworth inlier, Gloucestershire: Geol. Assoc. London Proc., v. 83, p. 1-35. (Southern Welsh Borderland)

Curtis, M.L.K., and Cave, R., 1964, The Silurian-Old Red Sandstone unconformity at Buckover, near Tortworth, Gloucestershire: Bristol Naturalists' Soc. Proc., v. 30, p. 427-442. (Southern Welsh Borderland)

Curtis, M.L.K., Lawson, J. D., Squirrell, H. C., Tucker, E. V., and Walmsley, V. G., 1967, The Silurian inliers of the southeastern Welsh Borderland: Geol. Assoc. London, Guide No. 5. (Southern Welsh Borderland)

Dairon, J., 1876, Notes on the Silurian rocks of Dumfriesshire, and their fossil remains: Glasgow Geol. Soc. Trans., v. 5, p. 176-184. (Southern Uplands)

——1879, On the rocks and graptolitic shales of the Moffat district: Glasgow Geol. Soc. Trans., v. 6, p. 178. (Southern Uplands)

Dakyns, J. R., 1869, Notes on the geology of the Lake District: Geol. Mag., dec. 1, v. 6, p. 56-58. (Northern England)

Dakyns, J. R., Tiddeman, R. H., and Goodchild, J. G., 1897, The geology of the country between Appleby, Ullswater, and Haweswater: Great Britain Geol. Survey Mem., sheet 102 S.W., old ser., 110 p. (Northern England)

Dana, J. D., 1890, Sedgwick and Murchison; Cambrian and Silurian: Am. Jour. Sci., ser. 3, v. 39, p. 167-180. (Correlation—General)

Das Gupta, T., 1932, The Salopian graptolite shales of the Long Mountain and similar rocks of Wenlock Edge: Geol. Assoc. London Proc., v. 43, p. 325-363. (Northern Welsh Borderland)

——1933, The zone of Monograptus vulgaris in the Welsh Borderland and North Wales: Liverpool Geol. Soc. Proc., v. 16, p. 109-115. (Wales and Welsh Borderland—General)

Davidson, T., and Maw, G., 1881, Notes on the physical character and thickness of the Upper Silurian rocks of Shropshire, with the brachiopoda they contain grouped in geological horizons: Geol. Mag., dec. 2, v. 8, p. 100-109. (Northern Welsh Borderland)

Davies, A. M., and Pringle, J., 1913, On two deep borings at Calvert Station (North Buckinghamshire) and on the Palaeozoic floor north of the Thames: Geol. Soc. London Quart. Jour., v. 69, p. 308-340. (Southeast England)

Davies, K. A., 1926, The geology of the country between Dryngarn and Abergwesyn, Breconshire: Geol. Soc. London Quart. Jour., v. 82, p. 436-464. (Central Wales)

——1928, Contributions to the geology of central Wales: Geol. Assoc. London Proc., v. 38, p. 157-168. (Central Wales)

——1933, The geology of the country between Abergwesyn (Breconshire) and Pumpsaint (Carmarthenshire): Geol. Soc. London Quart. Jour., v. 89, p. 172-200. (Central Wales)

Davies, K. A., and Platt, J. I., 1933, The conglomerates and grits of the Bala and Valentian rocks of the district between Rhayader (Radnorshire) and Llansawel (Carmarthenshire): Geol. Soc. London Quart. Jour., v. 89, p. 202-220. (Central Wales)

Davis, J. E., 1850, On the age and position of the limestone of Nash, near Presteigne, South Wales: Geol. Soc. London Quart. Jour., v. 6, p. 432-439. (Central Wales)

Dean, W. T., 1964, The geology of the Ordovician and adjacent strata in the southern Caradoc district of Shropshire: British Mus. (Nat. History) Bull., Geology, v. 1, p. 1-40. (Northern Welsh Borderland)

Dearman, W. R., Shiells, K.A.G., and Larwood, G. P., 1962, Refolded folds in the Silurian rocks of Eyemouth, Berwickshire: Yorkshire Geol. Soc. Proc., v. 33, p. 273-285. (Southern Uplands)

De La Beche, H. T., 1826, On the geology of southern Pembrokeshire: Geol. Soc. London Trans., ser. 2, v. 2, p. 1-20. (Southwest Wales)

——1846, On the formation of the rocks of South Wales and south-western England: Great Britain Geol. Survey Mem., v. 1, p. 1-296. (Wales and Welsh Borderland—General)

Dewey, J. F., 1963, The lower Palaeozoic stratigraphy of central Murrisk, County Mayo, Ireland, and the evolution of the South Mayo Trough: Geol. Soc. London Quart. Jour., v. 119, p. 313-344. (Ireland)

——1969, Evolution of the Appalachian/Caledonian orogen: Nature, v. 222, p. 124-129. (British Isles—General)

Dineley, D. L., 1950, The northern part of the lower Old Red Sandstone outcrop of the Welsh Borderland: Woolhope Naturalists' Field Club Trans., v. 33, p. 127-147. (Wales and Welsh Borderland—General)

——1960, Shropshire geology: An outline of the tectonic history: London, Field Studies, v. 1, 23 p. (Northern Welsh Borderland)

Dineley, D. L., and Gossage, D. W., 1959, The Old Red Sandstone of the Cleobury Mortimer area, Shropshire: Geol. Assoc. London Proc., v. 70, p. 221-238. (Northern Welsh Borderland)

Dines, H. G., Holmes, S.C.A., and Robbie, J. A., 1954, Geology of the country around Chatham: Great Britain Geol. Survey, sheet 272, new ser., 157 p. (Southeast England)

Dixon, E.E.L., 1921, Geology of the South Wales coalfield. Pt. XIII. The country around Pembroke and Tenby: Great Britain Geol. Survey Mem., sheets 244, 245, new ser., 220 p. (Southwest Wales)

Downie, C., 1963, 'Hystrichospheres' (Acritarchs) and spores of the Wenlock Shales (Silurian) of Wenlock, England: Palaeontology, v. 6, p. 625-652. (Correlations—General)

Drew, Helen, and Slater, I. L., 1910, Notes on the geology of the district around Llansawel (Carmarthenshire): Geol. Soc. London Quart. Jour., v. 66, p. 402-419. (Central Wales)

Earp, J. R., 1938, The higher Silurian rocks of the Kerry district, Montgomeryshire: Geol. Soc. London Quart. Jour., v. 94, p. 125-160. (Central Wales)

——1940, The geology of the south-western part of Clun Forest: Geol. Soc. London Quart. Jour., v. 96, p. 1-11. (Central Wales)

——1944, Observations on Upper Silurian graptolites: Geol. Mag., v. 81, p. 181-185. (Correlation—General)

Earp, J. R., and Hains, B. A., 1971, British regional geology: The Welsh Borderland (3d ed.): London, Her Majesty's Stationery Office, 118 p. (Wales and Welsh Borderland—General)

Eastwood, T., 1963, British regional geology: Northern England (3d ed): London, Her Majesty's Stationery Office, 71 p. (Northern England)

Eastwood, T., Whitehead, T. H., and Robertson, T., 1925, Geology of the country around Birmingham: Great Britain Geol. Survey Mem., sheet 168, new ser., 152 p. (Northern Welsh Borderland)

Eckford, R.J.A., and Mason, W., 1927, The geology of the district around Edinburgh: Dobb's Linn and Loch Skene. I. The Dobb's Linn Section: Geol. Assoc. London Proc., v. 38, p. 505-508. (Southern Uplands)

Eden, R. A., 1964, Scotland. 1. South Lowlands district, Geological Survey boreholes: Great Britain Geol. Survey Summary Progress for 1963, p. 52-54. (Midland Valley)

——1966, South Lowlands district; Eyemouth (34) sheet: Great Britain Geol. Survey Summary Progress for 1965, p. 61. (Southern Uplands)

Edmunds, F. H., 1923, The lower Ludlow rocks of the Clwŷdian Range, North Wales: Liverpool Geol. Soc. Proc., v. 13, p. 335-337. (North Wales)

Edmunds, F. H., and Stubblefield, C. J., 1936, On a bore reaching Palaeozoic strata at Bushey, Herts: Great Britain Geol. Survey Summary Progress for 1934, pt. 2, p. 32-34. (Southeast England)

Egan, F. W., and Bailey, W. H., 1872, Explanatory memoir to accompany sheet 48 of the maps of the Geological Survey of Ireland, illustrating parts of the counties of Down and Armagh: Geol. Survey Ireland Mem., 42 p. (Ireland)

——1873, Explanatory memoir to accompany sheet 47 of the maps of the Geological Survey of Ireland, including the country around Armagh: Geol. Survey Ireland Mem. (Ireland)

——1877, Explanatory memoir to accompany sheet 59 of the maps of the Geological Survey of Ireland, including the districts of Newtown, Hamilton, Keady, and Castleblayeny: Geol. Survey Ireland Mem., 32 p. (Ireland)

Elles, G. L., 1900, The zonal classification of the Wenlock Shales of the Welsh Borderland: Geol. Soc. London Quart. Jour., v. 56, p. 370-414. (Wales and Welsh Borderland—General)

——1909, The relation of the Ordovician and Silurian rocks of Conway, North Wales: Geol. Soc. London Quart. Jour., v. 65, p. 169-194. (North Wales)

——1922a, The Bala country: Its structure and rock-succession: Geol. Soc. London Quart. Jour., v. 78, p. 132-175. (North Wales)

——1922b, The graptolite faunas of the British Isles: Geol. Assoc. London Proc., v. 33, p. 168-200. (Correlation—General)

——1924, Evolutional palaeontology in relation to the lower Palaeozoic rocks: Rept. British Assoc. Adv. Sci. Liverpool, 1923, p. 83-107. (British Isles—General)

——1925, The characteristic assemblages of the graptolite zones of the British Isles: Geol. Mag., v. 62, p. 337-347. (Correlation—General)

——1939, Factors controlling graptolite succession and assemblages: Geol. Mag., v. 76, p. 181-187. (British Isles—General)

——1944, Upper Silurian graptolite zones: Geol. Mag., v. 81, p. 275-277. (Correlation—General)

Elles, G. L., and Slater, I. L., 1906, The highest Silurian rocks of the Ludlow district: Geol. Soc. London Quart. Jour., v. 62, p. 195-222. (Northern Welsh Borderland)

Elles, G. L., and Wood, E.M.R., 1896, On the Llandovery and associated rocks of Conway (North Wales): Geol. Soc. London Quart. Jour., v. 52, p. 273-288. (North Wales)

——1901-1918, A monograph of British graptolites: Paleontographical Soc. Mon., 539 p. (Correlation—General)

Etheridge, R., 1874, On the remains of *Pterygotus*, and other crustaceans from the Upper

Silurian Series of the Pentland Hills: Edinburgh Geol. Soc. Trans., v. 2, p. 314–316. (Midland Valley)

Etheridge, R., 1881, Anniversary address of the president: On the analysis and distribution of British Palaeozoic fossils: Geol. Soc. London Quart. Jour., v. 37, p. 51–235. (Southeast England)

Evans, D. C., 1906, The Ordovician rocks of western Carmarthenshire: Geol. Soc. London Quart. Jour., v. 62, p. 597–643. (Central Wales)

Fahraeus, L. E., 1969, Conodont zones in the Ludlovian of Gotland and a correlation with Great Britain: Sveriges Geol. Undersökning Årsb., v. 63, 33 p. (Correlation—General)

Falcon, N. L., 1947, Major clues in the tectonic history of the Malverns: Geol. Mag., v. 84, p. 229–240. (Southern Welsh Borderland)

——1952, The age of the Malvern folding: Geol. Mag., v. 89, p. 304. (Southern Welsh Borderland)

Falcon, N. L., and Kent, P. E., 1960, Geological results of petroleum exploration in Britain 1945–1957: Geol. Soc. London Mem. no. 2V, 56 p. (Southeast England)

Fearnsides, W. G., 1910, North and central Wales. Geology in the field (Jubilee volume): Geol. Assoc. London, p. 786–825. (Wales and Welsh Borderland—General)

Fearnsides, W. G., Elles, G. L., and Smith, B., 1907, The lower Palaeozoic rocks of Pomeroy: Royal Irish Acad. Proc., v. 26, p. 97–128. (Ireland)

Flett, J. S., and Hill, J. B., 1912, The geology of the Lizard and Meneage (sheet 359): Great Britain Geol. Survey Mem., sheet 359, new ser., 280 p. (Southwest England)

Forsyth, D., 1881, Notes on the Silurian rocks of the Muirkirk district: Glasgow Geol. Soc. Trans., v. 7, p. 74–77. (Midland Valley)

——1884, The Silurian rocks of the Girvan district: Glasgow Geol. Soc. Trans., v. 7, p. 358–369. (Midland Valley)

Foxall, W. H., 1917, The geology of the eastern boundary fault of the South Staffordshire coalfield: Birmingham Nat. History Philos. Soc. Proc., v. 14, p. 46–54. (Northern Welsh Borderland)

Freshney, E. C., 1960, An extension of the Silurian succession in the Craighead inlier, Girvan: Glasgow Geol. Soc. Trans., v. 24, p. 27–31. (Midland Valley)

Furness, R. R., 1965, The petrography and provenance of the Coniston Grits east of the Lune Valley, Westmorland: Geol. Mag., v. 102, p. 252–260. (Northern England)

Furness, R. R., Llewellyn, P. G., Norman, T. N., and Rickards, R. B., 1967, A review of Wenlock and Ludlow stratigraphy and sedimentation in N.W. England: Geol. Mag., v. 104, p. 132–147. (Northern England)

Gallois, R. W., and Edmunds, F. H., 1965, British regional geology: The Wealden district: London, Her Majesty's Stationery Office, 101 p. (Southeast England)

Gardiner, C. I., 1899, The Silurian and Ordovician rocks exposed on the shore near Balbriggan County Dublin: Geol. Mag., dec. 4, v. 6, p. 395–402. (Ireland)

——1916, The Silurian inlier of Usk: Cotteswold Naturalists' Field Club Proc., v. 19, p. 129–172. (Southern Welsh Borderland)

——1920, The Silurian rocks of May Hill: Cotteswold Naturalists' Field Club Proc., v. 20, p. 185–222. (Southern Welsh Borderland)

——1927, The Silurian inlier of Woolhope (Herefordshire): Geol. Soc. London Quart. Jour., v. 83, p. 501–530. (Southern Welsh Borderland)

——1937, A roadside section two miles west of Huntley: Cotteswold Naturalists' Field Club, v. 26, p. 169–172. (Southern Welsh Borderland)

Gardiner, C. I., and Reynolds, S. H., 1897, An account of the Portraine inlier, County Dublin: Geol. Soc. London Quart. Jour., v. 53, p. 520–539. (Ireland)

——1902, The fossiliferous Silurian beds and the associated igneous rocks of the Clogher Head district: Geol. Soc. London Quart. Jour., v. 57, p. 226–266. (Ireland)

——1912, Ordovician and Silurian rocks of Kilbride Peninsula, County Mayo: Geol. Soc. London Quart. Jour., v. 68, p. 75-102. (Ireland)

——1914, The Ordovician and Silurian rocks of the Lough Nafooey area (County Galway): Geol. Soc. London Quart. Jour., v. 70, p. 104-118. (Ireland)

Gardiner, C. I., Reynolds, S. H., Smith, S., Trueman, A. E., and Tutcher, J. W., 1934, The geology of the Gloucester district: Geol. Assoc. London Proc., v. 45, p. 109-144. (Southern Welsh Borderland)

Garwood, E. J., and Goodyear, E., 1918, On the geology of the Old Radnor district, with special reference to an algal development in the Woolhope Limestone: Geol. Soc. London Quart. Jour., v. 74, p. 1-30. (Central Wales)

Geikie, A., 1864, The geology of eastern Berwickshire: Great Britain Geol. Survey Mem., sheet 34, p. 1-58. (Southern Uplands)

——1871, On the order and succession among the Silurian rocks of Scotland: Glasgow Geol. Soc. Trans., v. 3, p. 74-95. (Scotland—General)

——1873, The Southern Uplands of Scotland: Nature, v. 9, p. 81-82. (Southern Uplands)

——1897, Ancient volcanoes of Great Britain, Vol. I: London, MacMillan and Co., 477 p. (British Isles—General)

George, T. N., 1960, The stratigraphical evolution of the Midland Valley: Glasgow Geol. Soc. Trans., v. 24, p. 32-107. (Midland Valley)

——1961, British regional geology: North Wales: London, Her Majesty's Stationery Office, 96 p. (North Wales)

——1962, Tectonics and paleogeography in southern England: Sci. Progress, v. 50, p. 192-217. (Southeast England)

——1963, Palaeozoic growth of the British Caledonides, in Johnson, M.R.W., and Stewart, F. H., eds., The British Caledonides: Edinburgh and London, Oliver & Boyd, p. 1-33. (British Isles—General)

——1970, British regional geology: South Wales: London, Her Majesty's Stationery Office, 152 p. (Southwest Wales)

George, T. N., Cocks, L.R.M., and Toghill, P., 1971, Stages of the Llandovery Series: Geol. Mag., v. 108, p. 263-265. (Correlation—General)

——1972, Stages of the Llandovery Series: Geol. Mag., v. 109, p. 67-68. (Correlation—General)

Glass, N., 1861, Silurian strata near Cardiff: Geologist, v. 4, p. 168. (Southern Welsh Borderland)

Gordon, A. J., 1962, The Lower Palaeozoic rocks around Glenluce, Wigtownshire [Ph.D. thesis]: Edinburgh, Edinburgh Univ. (Southern Uplands)

Green, D. H., 1966, A re-study and re-interpretation of the geology of the Lizard peninsula, Cornwall, in Hosking, K.F.G., and Shrimpton, C. J., eds., Present views of some aspects of the geology of Cornwall and Devon: Royal Geol. Soc. Cornwall, p. 87-114. (Southwest England)

Green, G. W., and Welch, F.B.A., 1965, Geology of the country around Wells and Cheddar: Great Britain Geol. Survey Mem., sheet 280, new ser., 225 p. (Southern Welsh Borderland)

Green, J.F.N., 1912, The older Palaeozoic succession of the Duddon Estuary: London, Hayman, Christy and Lilly, Ltd., 23 p. (Northern England)

——1915, The structure of the eastern part of the Lake District: Geol. Assoc. London Proc., v. 26, p. 195-223. (Northern England)

——1920, The geological structure of the Lake District: Geol. Assoc. London Proc., v. 31, p. 109-126. (Northern England)

Green, U., 1904, On the discovery of Silurian fossils of Ludlow age in Cornwall: Geol. Mag., dec. 5, v. 1, p. 289-290. (Southwest England)

Green, U., and Sherborn, C. D., 1906, Lists of Wenlockian fossils from Porthluney, Cornwall; Ludlovian fossils from Porthalla; and Taunusian fossils from Polyne Quarry, near Looe, Cornwall: Geol. Mag., dec. 5, v. 3, p. 33-35. (Southwest England)

Greenly, Edward, 1919, The geology of Anglesey: Great Britain Geol. Survey Mem., 980 p. (North Wales)

Greig, D. C., Wright, J. E., Hains, B. A., and Mitchell, G. H., 1968, Geology of the country around Church Stretton, Craven Arms, Wenlock Edge, and Brown Clee: Great Britain Geol. Survey Mem., sheet 166, new ser., 379 p. (Northern Welsh Borderland)

Griffith, A. E., 1961, A note on some shelly fossils from the arenaceous greywackes of County Down: Irish Naturalists' Jour., v. 13, p. 258-259. (Ireland)

Griffith, R. J., 1844, On the Old Red Sandstone, or Devonian and Silurian districts of Ireland: Rept. British Assoc. Adv. Sci., Cork, 1843, Trans. Secs., p. 46-49. (Ireland)

Groom, T. T., 1899, The geological structure of the Southern Malverns and of the adjacent district to the west: Geol. Soc. London Quart. Jour., v. 55, p. 129-169. (Southern Welsh Borderland)

——1900, On the geological structure of portions of the Malvern and Abberley Hills: Geol. Soc. London Quart. Jour., v. 56, p. 138-197. (Southern Welsh Borderland)

——1910, The geology of the Malvern and Abberley Hills, and the Ledbury district (Jubilee volume): Geol. Assoc. London, p. 698-738. (Southern Welsh Borderland)

Groom, T. T., and Lake, P., 1908, The Bala and Llandovery rocks of Glyn Ceiriog, North Wales: Geol. Soc. London Quart. Jour., v. 64, p. 546-595. (North Wales)

Gunn, W., and Clough, C. T., 1878, Discovery of Silurian beds in Teesdale: Geol. Soc. London Quart. Jour., v. 24, p. 27-34. (Northern England)

Hains, B. A., 1970, The geology of the Wenlock Edge area (explanation of 1:25,000 geol. sheet S059): Great Britain Geol. Survey, 61 p. (Northern Welsh Borderland)

Hains, B. A., and Horton, A., 1969, British regional geology: Central England: London, Her Majesty's Stationery Office, 142 p. (Northern Welsh Borderland)

Hallissy, T., 1914, Clare Island survey. Pt. 7. Geology: Royal Irish Acad. Proc., v. 31B, p. 1-22. (Ireland)

Hardie, W. G., 1954, The Silurian rocks of Kendal End, near Barnt Green, Worcestershire: Geol. Assoc. London Proc., v. 65, p. 11-17. (Northern Welsh Borderland)

Harkness, R., 1851, On the Silurian rocks of Dumfriesshire and Kirkurdbrightshire: Geol. Soc. London Quart. Jour., v. 7, p. 46-58 (Southern Uplands)

——1853, On the Silurian rocks of Kirkcudbrightshire: Geol. Soc. London Quart. Jour., v. 9, p. 181-186. (Southern Uplands)

——1856, On the lowest sedimentary rocks in the south of Scotland: Geol. Soc. London Quart. Jour., v. 12, p. 238-245. (Southern Uplands)

——1873, The Southern Uplands of Scotland: Nature, v. 9, p. 22-24 and 57-59. (Southern Uplands)

Harkness, R., and Nicholson, H. A., 1866, Additional observations on the geology of the Lake Country (with a note on two new species of trilobites by J. W. Salter): Geol. Soc. London Quart. Jour., v. 22, p. 480-486. (Northern England)

——1868, On the Coniston Group: Geol. Soc. London Quart. Jour., v. 24, p. 296-303. (Northern England)

——1877, On the strata and their fossil contents between the Borrowdale Series of the north of England and the Coniston Flags: Geol. Soc. London Quart. Jour., v. 33, p. 461-484. (Northern England)

Harland, W. B. and Gayer, R. A., 1972, The Arctic Caledonides and earlier oceans: Geol. Mag., v. 109, p. 289-314. (British Isles—General)

Harper, J. C., 1948, The Ordovician and Silurian rocks of Ireland: Liverpool Geol. Soc. Proc., v. 20, p. 48-67. (Ireland)

Harper, J. C., and Brenchley, P. J., 1972, Some points of interest concerning the Silurian inliers of southwest central Ireland in their geosynclinal context: A statement: Geol. Soc. London Jour., v. 128, p. 257-262. (Ireland)

Harper, J. C., and Hartley, J. J., 1938, The Silurian inlier of Lisbellaw, County Fermanagh, with a note on the age of the Fintona Beds: Royal Irish Acad. Proc., v. 45, sec. B, p. 73-87. (Ireland)

Hartley, J. J., 1933, The geology of north-east Tyrone and the adjacent portions of County Londonderry: Royal Irish Acad. Proc., v. 41, sec. B, p. 218-255. (Ireland)

Haswell, G. C., 1865, On the Silurian formation in the Pentland Hills: Edinburgh Geol. Soc., 45 p. (Midland Valley)

Hawkins, H. L., 1942, Some episodes in the geological history of the south of England: Geol. Soc. London Quart. Jour., v. 98, p. xlix-lxx. (Southeast England)

Hawkins, T.R.W., 1966, Boreholes at Parys Mountain, near Amlwch, Anglesey: Great Britain Geol. Survey Bull., no. 24, p. 7-18. (North Wales)

Heard, A., and Davies, R., 1924, The Old Red Sandstone of the Cardiff district: Geol. Soc. London Quart. Jour., v. 80, p. 489-519. (Southern Welsh Borderland)

Henderson, J., and Brown, D. J., 1869, On the Silurian beds of the Pentland Hills: Geol. Mag., dec. 1, v. 6, p. 228-229. (Midland Valley)

——1870, On the Silurian rocks of the Pentland Hills. Pt. II: Edinburgh Geol. Soc. Trans., v. 1, p. 266-272. (Midland Valley)

Hendriks, E.M.L., 1926, The Bala-Silurian succession in the Llangranog district (South Cardiganshire): Geol. Mag., v. 63, p. 121-139. (Central Wales)

——1937, Rock succession and structure in south Cornwall: A revision. With notes on the central European facies and Variscan folding there present: Geol. Soc. London Quart. Jour., v. 93, p. 322-367. (Southwest England)

Hicks, Henry, 1872, On the classification of the Cambrian and Silurian rocks: Geol. Mag., dec. 1, v. 9, p. 383-384. (Correlation—General)

——1873, On the classification of the Cambrian and Silurian rocks: Geol. Assoc. London Proc., v. 3, p. 99. (Correlation—General)

——1876a, Some considerations on the probable conditions under which the Palaeozoic rocks were deposited over the Northern Hemisphere: Geol. Mag., dec. 2, v. 3, p. 156-160, 215-218, and 249-253. (British Isles—General)

——1876b, Llandovery rocks in the Lake District: Geol. Mag., dec. 2, v. 3, p. 335-336 and 429-430. (Northern England)

——1876c, On some areas where the Cambrian and Silurian rocks occur as a conformable series: Rept. British Assoc. Adv. Sci., Bristol, 1875, Trans. Secs., p. 69. (British Isles—General)

——1879, The classification of the Eozoic and lower Palaeozoic rocks of the British Isles: Popular Sci. Rev., ser. 2, v. 5, p. 289. (Correlation—General)

——1881, On the discovery of plants at the base of the Denbighshire Grits, near Corwen, North Wales: Geol. Soc. London Quart. Jour., v. 37, p. 482-496. (North Wales)

Hill, D., 1936, Report of "Coral Reef" meeting at Wenlock Edge, the Dudley district, and the Oxford district; Wenlock Edge: Geol. Assoc. London Proc., v. 47, p. 130-133. (Northern Welsh Borderland)

Hinde, G. J., 1904, The Bone Bed in the Upper Ludlow Formation: Geol. Assoc. London Proc., v. 18, p. 443-446. (Wales and Welsh Borderland—General)

Holl, H. B., 1865, On the geological structure of the Malvern Hills and adjacent districts: Geol. Soc. London Quart. Jour., v. 21, p. 72-102. (Southern Welsh Borderland)

Holland, C. H., 1958, The Ludlovian and Downtonian rocks of the Knighton district, Radnorshire: Geol. Soc. London Proc., no. 1558, p. 45. (Central Wales)

——1959a, On convolute bedding in the lower Ludlovian rocks of northeast Radnorshire: Geol. Mag., v. 96, p. 230-236. (Central Wales)

——1959b, The Ludlovian and Downtonian rocks of the Knighton district, Radnorshire: Geol. Soc. London Quart. Jour., v. 114, p. 449-482. (Central Wales)

Holland, C. H., 1962, The Ludlovian-Downtonian succession in central Wales and the central Welsh Borderland: Symposium-Band der 2. internationalen Arbeitstagung über die Silur Devon-Grenze und die Stratigraphie von Silur und Devon: Stuttgart, E. Schweizerbart'sche Verlag, p. 87-94. (Wales and Welsh Borderland—General)

——1965, The Siluro-Devonian boundary: Geol. Mag., v. 102, p. 213-221. (Correlation—General)

——1969a, Irish counterpart of Silurian of Newfoundland, in Kay, M., ed., North Atlantic geology and continental drift: Am. Assoc. Petroleum Geologists Mem. 12, p. 298-308. (Ireland)

——1969b, The Welsh Silurian geosyncline in its regional context, in Wood, A., ed., The Pre-Cambrian and lower Palaeozoic rocks of Wales: Cardiff, Wales Univ. Press, p. 203-217. (Wales and Welsh Borderland—General)

Holland, C. H., and Lawson, J. D., 1963, Facies patterns in the Ludlovian of Wales and the Welsh Borderland: Geol. Jour., v. 3, p. 269-288. (Wales and Welsh Borderland—General)

Holland, C. H., Lawson, J. D., and Walmsley, V. G., 1959, A revised classification of the Ludlovian succession at Ludlow: Nature, v. 184, p. 1037-1039. (Northern Welsh Borderland)

——1962, Ludlovian classification—A reply: Geol. Mag., v. 99, p. 393-398. (Correlation—General)

——1963, The Silurian rocks of the Ludlow district, Shropshire: British Mus. (Nat. History) Bull., Geology, p. 95-171. (Northern Welsh Borderland)

Holland, C. H., Rickards, R. B., and Warren, P. T., 1969, The Wenlock graptolites of the Ludlow district, Shropshire, and their stratigraphical significance: Palaeontology, v. 12, p. 663-683. (Northern Welsh Borderland)

Hollingworth, S. E., 1954, The geology of the Lake District—A review (with contributions by W.C.C. Rose, R. L. Oliver, and R. J. Firman): Geol. Assoc. London Proc., v. 65, p. 385-402. (Northern England)

Holtedahl, O., 1939, Correlation notes on Scottish-Norwegian Caledonian geology: Norsk Geol. Tidsskr., v. 19, p. 326-339. (British Isles—General)

Hopkins, W., 1853, On the geological structure of the Palaeozoic rocks of Wales and the adjoining English countries: Geol. Soc. London Quart. Jour., v. 9, p. lxi-xcii. (Wales and Welsh Borderland—General)

Hopkinson, J., 1880, On the recent discovery of Silurian rocks in Hertfordshire and their relation to the water-bearing strata of the London basin: Watford Nat. History Soc. Trans., v. 2, p. 241-248. (Southeast England)

House, M. R., and Sellwood, E. B., 1966, Palaeozoic paleontology in Devon and Cornwall, in Hosking, K.F.G., and Shrimpton, G. J., eds., Present views on some aspects of the geology of Cornwall and Devon: Royal Geol. Soc. Cornwall, p. 45-86. (Southwest England)

Howard, F. T., and Small, E. W., 1894, On some igneous rocks of south Pembrokeshire with a note on the rocks of the Isle of Grassholme: Rept. British Assoc. Adv. Sci., Nottingham, 1893, Trans. Secs., p. 766-767. (Southwest Wales)

——1896a, The geology of Skomer Island: Rept. British Assoc. Adv. Sci., Liverpool, 1896, Trans. Secs., p. 797-798. (Southwest Wales)

——1896b, Geological notes on Skomer Island: Cardiff Nat. Soc. Trans., v. 28, p. 55-60. (Southwest Wales)

——1897, Further notes on Skomer Island: Cardiff Nat. Soc. Trans., v. 29, p. 62-63. (Southwest Wales)

Howell, H. H., and Geikie, A., 1861, The geology of the neighborhood of Edinburgh: Great Britain Geol. Survey Mem., sheet 32, 151 p. (Midland Valley)

Hubert, J. F., Scott, K. M., and Walton, E. K., 1966, Composite nature of Silurian flysch sandstones shown by groove moulds on intro-bed surfaces, Peebleshire, Scotland: Jour. Sed. Petrology, v. 36, p. 237-241. (Southern Uplands)

Hughes, T. McK., 1876, Notes on the classification of the sedimentary rocks: Rept. British Assoc. Adv. Sci., Bristol, 1875, Trans. Secs., p. 70-73. (Correlation—General)

——1879, On the Silurian rocks of the Valley of the Clwŷd: Geol. Soc. London Quart. Jour., v. 35, p. 694-698. (North Wales)

——1885, Notes on the geology of the Vale of Clwŷd: Chester Soc. Nat. Sci. Proc., v. 3, p. 5-37. (North Wales)

——1894, Observations on the Silurian rocks of North Wales: Chester Soc. Nat. Sci. Proc., v. 4, p. 144-160. (North Wales)

——1902, Ingleborough. Pt. II. Stratigraphy: Yorkshire Geol. Soc. Proc., v. 14, p. 323-343.(Northern England)

——1906, Ingleborough. Pt. II(III). Stratigraphy: Yorkshire Geol. Soc. Proc., v. 15, p. 351-371. (Northern England)

——1907, Ingleborough. Pt. IV. Stratigraphy and palaeontology of the Silurian: Yorkshire Geol. Soc. Proc., v. 16, p. 45-74. (Northern England)

Hull, E., 1879, On the geological age of the rocks forming the southern highlands of Ireland, generally known as "The Dingle Beds" and Glengariff Grits and Slates (Jukes): Geol. Soc. London Quart. Jour., v. 35, p. 699-723. (Ireland)

——1882, On a proposed Devono-Silurian formation: Geol. Soc. London Quart. Jour., v. 38, p. 200-209. (Correlation—General)

——1888, Explanatory memoir to accompany sheets 138 and 139 of the map of the Geological Survey of Ireland, with an account of the mines, by R. J. Cruise and notes on the igneous rocks by F. H. Hatch: Geol. Survey Ireland Mem., 55 p. (Ireland)

Hunt, T. S., 1873, History of the names Cambrian and Silurian in geology: Geol. Mag., dec. 1, v. 10, p. 385-395 and 453-461. (Correlation—General)

Hunter, J.R.S., 1881, The Silurian rocks of Logan Water, Lesmahagow: Glasgow Geol. Soc. Trans., v. 7, p. 56-64. (Midland Valley)

——1882, The geology and palaeontology of Bankend, Bellfield, and Coalburn, Lesmahagow: Glasgow Geol. Soc. Trans., v. 7, p. 143-157. (Midland Valley)

——1884, The Silurian districts of Leadhills and Wanlockhead, and their early and recent mining history: Glasgow Geol. Soc. Trans., v. 7, p. 373-392. (Southern Uplands)

Hutt, J. E., 1973, Lake District Llandovery graptolites [Ph.D. thesis]: Leicester, Leicester Univ. (Northern England)

James, D.M.G., and James, J., 1969, The influence of deep fractures on some areas of Ashgillian-Llandoverian sedimentation in Wales: Geol. Mag., v. 106, p. 562-582. (Central Wales)

Jehu, R. M., 1926, The geology of the district around Towyn and Abergynolwyn, Merioneth: Geol. Soc. London Quart. Jour., v. 82, p. 465-489. (North Wales)

Jennings, J. S., 1961, The geology of the eastern part of the Lesmagow inlier [Ph.D. thesis]: Edinburgh, Edinburgh Univ. (Midland Valley)

Jones, O. T., 1909, The Hartfell-Valentian succession in the district around Plynlimon and Pont Erwyd, North Cardiganshire: Geol. Soc. London Quart. Jour., v. 65, P. 463-537. (Central Wales)

——1912, The geological structure of central Wales and the adjoining regions: Geol. Soc. London Quart. Jour., v. 68, p. 328-344. (Central Wales)

——1918, Gothlandian or Silurian. a. Great Britain, in Evans, J. W., ed., Geology of the British Isles: The Hague, Martins Nijhoff, p. 82-102. (British Isles—General)

——1921, The Valentian Series: Geol. Soc. London Quart. Jour., v. 77, p. 144-174. (Correlation—General)

Jones, O. T., 1922, The mining district of north Cardiganshire and west Montgomeryshire: Great Britain Geol. Survey Mem., Mineral Resources, v. 20, 207 p. (Central Wales)

———1923, The Hirnant beds and the base of the Valentian: Geol. Mag., v. 60, p. 514-519. (North Wales)

———1924, The age of the Hirnant Beds: Geol. Mag., v. 61, p. 283-284. (North Wales)

———1925a, The Llandovery rocks of Llandovery: Rept. British Assoc. Adv. Sci., Toronto, 1924, Trans. Secs., p. 394. (Central Wales)

———1925b, Geology of the Llandovery district. Pt. I: The southern area: Geol. Soc. London Quart. Jour., v. 81, p. 344-388. (Central Wales)

———1925c, The Ordovician-Silurian boundary in Britain and North America: Jour. Geology, v. 33, p. 371-388. (British Isles—General)

———1927, The foundations of the Pennines: Geol. Jour., v. 1, p. 5-14. (Northern England)

———1929, Silurian, in Evans, J. W., and Stubblefield, C. J., eds., Handbook of the geology of Great Britain: London, Thomas Murby and Co., p. 88-127. (British Isles—General)

———1935, The lower Palaeozoic rocks of Britain: Internat. Geol. Cong., 16th, Washington, D.C. 1933, Rept., v. 1, p. 463-484. (British Isles—General)

———1937, On the sliding or slumping of submarine sediments in Denbighshire, North Wales, during the lower Ludlow Period: Geol. Soc. London Quart. Jour., v. 93, p. 241-283. (North Wales)

———1938, On the evolution of a geosyncline (presidential address): Geol. Soc. London Quart. Jour., v. 94, p. lx-cx. (British Isles—General)

———1939, The geology of the Colwyn Bay district: A study of submarine slumping during the Salopian Period: Geol. Soc. London Quart. Jour., v. 95, p. 335-381. (North Wales)

———1943, A comment on a new area of slumped beds in Denbighshire: Geol. Mag., v. 80, p. 66-68. (North Wales)

———1947a, The geology of the Silurian rocks west and south of the Carneddau Range, Radnorshire: Geol. Soc. London Quart. Jour., v. 103, p. 1-36. (Central Wales)

———1947b, The Llandoverian graptolite succession in Britain: Mus. Royal Histoire Nat. Belge, ser. B., v. 23, no. 22, p. 1-3. (Correlation—General)

———1949, Geology of the Llandovery district. Pt. 2: Geol. Soc. London Quart. Jour., v. 105, p. 43-64. (Central Wales)

———1953, On submarine slumping in the lower Ludlow rocks of North Wales: Geol. Mag., v. 90, p. 220-221. (North Wales)

———1956, The geological evolution of Wales and the adjacent regions: Geol. Soc. London Quart. Jour., v. 111, p. 323-351. (Wales and Welsh Borderland—General)

Jones, O. T., and Pugh, W. J., 1915, The geology of the district around Machynlleth and the Llyfnant Valley: Geol. Soc. London Quart. Jour., v. 71, p. 343-385. (Central Wales)

———1935a, The geology of the districts around Machynlleth and Aberystwyth (Wales): Geol. Assoc. London Proc., v. 46, p. 247-300. (Central Wales)

———1935b, Summer field meeting to the Aberystwyth district (Wales): Geol. Assoc. London Proc., v. 46, p. 413-428. (Central Wales)

Jones, R. K., Brooks, M., Bassett, M. G., Austin, R. L., and Aldridge, R. J., 1969, An upper Llandovery limestone overlying Hollybush Sandstone (Cambrian) in Hollybush Quarry, Malvern Hills: Geol. Mag., v. 106, p. 457-469. (Southern Welsh Borderland)

Jones, W.D.V., 1945, The Valentian succession around Llanidloes, Montgomeryshire: Geol. Soc. London Quart. Jour., v. 100, p. 309-332. (Central Wales)

Jukes, J. B., 1853, On the occurrence of Caradoc sandstone at Great Barr, south Staffordshire: Geol. Soc. London Quart. Jour., v. 9, p. 179-181. (Northern Welsh Borderland)

Jukes, J. B., and Du Noyer, G. V., 1860, On the geological structure of Cahercohree Mountain, ten miles west of Tralee: Geol. Soc. Dublin Jour., v. 8, p. 106-109. (Ireland)

Jukes, J. B., Kinahan, G. H., and Wynne, A. B., 1860, Explanations to accompany sheet

144, of the maps of the Geological Survey of Ireland, illustrating parts of the counties of Limerick, Tipperary, and Clare: Geol. Survey Ireland Mem., 38 p. (Ireland)

Jukes-Browne, A. J., 1911, The building of the British Isles (3d ed.): London, Edward Stanford, 470 p. (British Isles—General)

Keeping, W., 1878, Notes on the geology of the neighborhood of Aberystwyth: Geol. Mag., dec. 2, v. 5, p. 532-547. (Central Wales)

——1881, The geology of central Wales: Geol. Soc. London Quart. Jour., v. 37, p. 141-170. (Central Wales)

Kellaway, G. A., 1963, Midlands, and central and North Wales district, Geological Survey boreholes: Great Britain Geol. Survey Summary Progress for 1962, p. 38-40. (Northern Welsh Borderland)

——1964, Atherstone (155) sheet: Great Britain Geol. Survey Summary Progress for 1963, p. 42-43. (Northern Welsh Borderland)

——1966, South Midlands, and central Wales district; Aberystwyth (163) sheets: Great Britain Geol. Survey Summary Progress for 1965, p. 48. (Central Wales)

Kellaway, G. A., and Welch, F.B.A., 1948, British regional geology: Bristol and Gloucester district (2d ed.): London, Her Majesty's Stationery Office, 91 p. (Southern Welsh Borderland)

Kelling, G., 1961, The stratigraphy and structure of the Ordovician rocks of the Rhinns of Galloway: Geol. Soc. London Quart. Jour., v. 117, p. 37-75. (Southern Uplands)

——1964, The turbidite concept in Britain: Turbidites, in Bouma, A. H., and Brouwer, A., eds., Developments in sedimentation—3: Amsterdam, Elsevier Pub. Co., p. 75-92. (British Isles—General)

Kelling, G., and Woolands, M. A., 1969, The stratigraphy and sedimentation of the Llandoverian rocks of the Rhayader district, in Wood, A., ed., The Pre-Cambrian and lower Palaeozoic rocks of Wales: Cardiff, Wales Univ. Press, p. 255-282. (Central Wales)

Kelly, J., 1860, Graywacke rocks of Ireland: Geol. Soc. Dublin Jour., v. 8, p. 251-333. (Ireland)

Kendall, P. F., and Wroot, H. E., 1924, Geology of Yorkshire: Vienna, Hollinek Bros., 995 p. (Northern England)

Kennedy, W. Q., 1958, The tectonic evolution of the Midland Valley of Scotland: Glasgow Geol. Soc. Trans., v. 23, p. 106-133. (Midland Valley)

Kilroe, J. R., 1907, The Silurian and metamorphic rocks of Mayo and north Galway: Royal Irish Acad. Proc., v. 26, p. 129-160. (Ireland)

Kilroe, J. R., and McHenry, A., 1901, The relation of the Silurian rocks of Ireland to the great metamorphic series: Rept. British Assoc. Adv. Sci., Glasgow, 1901, Trans. Secs., p. 636-637. (Ireland)

Kinahan, G. H., 1874, Geology of west Galway and southwest Mayo: Geol. Mag., dec. 2, v. 1, p. 453-462. (Ireland)

——1887a, A table of Irish lower Palaeozoic rocks, with their probable English equivalents: Royal Geol. Soc. Ireland Jour., v. 7, p. 98-103. (Ireland)

——1887b, The Lisbellaw Conglomerate, County Fermanagh, and Chesil Bank, Dorsetshire: Royal Geol. Soc. Ireland Jour., v. 7, p. 191-193. (Ireland)

Kinahan, G. H., and Bailey, W. H., 1883, Report on the rocks of the Fintona and Curlew Mountain district: Royal Irish Acad. Proc., v. 3, 2d ser., p. 475-500. (Ireland)

Kinahan, G. H., and others, 1876, Memoir to accompany sheets 83 and 84: Geol. Survey Ireland Mem., 84 p. (Ireland)

——1878, Memoir to accompany sheets 93 and 94: Geol. Survey Ireland Mem., 177 p. (Ireland)

King, W.B.R., 1923, The Upper Ordovician rocks of the southwestern Berwyn Hills: Geol. Soc. London Quart. Jour., v. 79, p. 487-507. (North Wales)

King, W.B.R., 1928, The geology of the district around Meifod, Montgomeryshire: Geol. Soc. London Quart. Jour., v. 84, p. 671-702. (Central Wales)

King, W.B.R., and Wilcockson, W. H., 1934, The lower Palaeozoic rocks of Austwick and Horton-in-Ribblesdale: Geol. Soc. London Quart. Jour., v. 90, p. 7-31. (Northern England)

King, W. W., 1925, Notes on the "Old Red Sandstone" of Shropshire: Geol. Assoc. London Proc., v. 36, p. 383-389. (Northern Welsh Borderland)

——1934, The Downtonian and Dittonian strata of Great Britain and north-western Europe: Geol. Soc. London Quart. Jour., v. 90, p. 526-570. (British Isles—General)

King, W. W., and Lewis, W. J., 1912, The uppermost Silurian and Old Red Sandstone of south Staffordshire: Geol. Mag., dec. 5, v. 9, p. 437-443 and 484-491. (Northern Welsh Borderland)

——1917, The Downtonian of south Staffordshire: Birmingham Nat. History Philos. Soc. Proc., v. 14, p. 90-99. (Northern Welsh Borderland)

Kirk, N. H., 1951a, The upper Llandovery and lower Wenlock rocks of the area between Dolyhir and Presteigne, Radnorshire: Geol. Soc. London Proc., no. 1471, p. 56-58. (Central Wales)

——1951b, The Silurian and Downtonian rocks of the anticlinal disturbance of Breconshire and Radnorshire: Pont Faen to Presteigne: Geol. Soc. London Proc., no. 1474, p. 72-74. (Central Wales)

Knill, J. L., 1959, Axial and marginal sedimentation in geosynclinal basins: Jour. Sed. Petrology, v. 29, p. 317-325. (British Isles—General)

Kuenen, P. H., 1953, Graded bedding, with observations on lower Palaeozoic rocks of Britain: Koninkl. Nederlandse Akad. Wetensch. Verh., Afd. Natuurk., v. 20, p. 1-47. (British Isles—General)

Laird, M. G., 1968, Rotational slumps and slump scars in Silurian rocks, western Ireland: Sedimentology, v. 10, p. 111-120. (Ireland)

Laird, M. G., and McKerrow, W. S., 1970, The Wenlock sediments of north-west Galway, Ireland: Geol. Mag., v. 107, p. 297-317. (Ireland)

Lake, P., 1895, The Denbighshire Series of south Denbighshire: Geol. Soc. London Quart. Jour., v. 51, p. 9-23. (North Wales)

Lake, P., and Groom, T. T., 1893, The Llandovery and associated rocks of the neighbourhood of Corwen: Geol. Soc. London Quart. Jour., v. 49, p. 426-440. (North Wales)

Lambert, J.L.M., 1965, The unstratified sedimentary rocks of the Meneage area, Cornwall: Geol. Soc. London Proc., v. 1619, p. 17-20. (Southwest England)

Lamont, A., 1935, The Drummuck Group, Girvan: A stratigraphical revision, with descriptions of new fossils from the lower part of the group: Glasgow Geol. Soc. Trans., v. 19, p. 289-330. (Midland Valley)

——1936, Palaeozoic seismicity: Nature, v. 138, p. 243-244. (British Isles—General)

——1940, Derived upper Llandovery fossils in Bunter pebbles from near Cheadle, north Staffordshire: Cement, Lime, and Gravel, v. 15, p. 26-30. (Northern Welsh Borderland)

——1945, Excursion to Onny River, Shropshire: London, Quarry Managers' Jour., v. 29, p. 118-119. (Northern Welsh Borderland)

——1946, Fossils from Middle Bunter Pebbles collected in Birmingham: Geol. Mag., v. 83, p. 39-44. (Northern Welsh Borderland)

——1947, Gala-Tarannon Beds in the Pentland Hills, near Edinburgh: Geol. Mag., v. 84, p. 193-208 and 289-303. (Midland Valley)

——1952, Ecology and correlation of the Pentlandian—A new division of the Silurian System in Scotland: Internat. Geol. Cong., 18th, London 1948, Rept., pt. 10, p. 27-32. (Midland Valley)

——1965, Varieties of *Conchidium knightii* (J. Sowerby) from the Wenlock and Aymestry Limestones: Scottish Jour. Sci., v. 1, p. 19-32. (British Isles—General)

Lamont, A., and Gilbert, D.L.F., 1945, Upper Llandovery Brachiopoda from Coneygore Coppice and Old Storridge Common, near Alfrick, Worcs.: Ann. Mag. Nat. History, ser. 11, v. 12, p. 641-682. (Southern Welsh Borderland)

Lamplugh, G. W., and Kitchen, F. L., 1911, Mesozoic rocks in some of the coal explorations in Kent: Great Britain Geol. Survey Mem., 212 p. (Southeast England)

Lamplugh, G. W., Wilkinson, S. B., Kilroe, J. R., McHenry, A., Seymour, H. J., and Wright, W. B., 1907, The geology of the country around Limerick: Geol. Survey Ireland Mem., 119 p. (Ireland)

Lamplugh, G. W., Kitchen, F. L., and Pringle, J., 1923, The concealed Mesozoic rocks of Kent: Great Britain Geol. Survey Mem., 248 p. (Southeast England)

Lapworth, C., 1870, On the Lower Silurian rocks of Galashiels: Geol. Mag., dec. 1, v. 7, p. 204-209 and 279-284. (Southern Uplands)

——1872a, On the Silurian rocks of the south of Scotland: Glasgow Geol. Soc. Trans., v. 4, p. 164-174. (Southern Uplands)

——1872b, On the graptolites of the Gala Group: Rept. British Assoc. Adv. Sci., Edinburgh, 1871, Trans. Secs., p. 104. (Southern Uplands)

——1872c, Note on the result of some recent researches among the graptolitic black shales of the south of Scotland: Geol. Mag., dec. 1, v. 9, p. 533-535. (Southern Uplands)

——1874a, Note on the graptolites discovered by John Henderson in the Silurian shales of Habbies Howe, Pentland Hills: Edinburgh Geol. Soc. Trans., v. 2, p. 375-377. (Midland Valley)

——1874b, On the Silurian rocks of the south of Scotland: Glasgow Geol. Soc. Trans., v. 4, p. 164-174. (Southern Uplands)

——1876a, Silurian rocks of the west of Scotland (with figures of the graptolites): Catalogue of the western Scottish fossils: Glasgow, British Assoc. Adv. Sci., 28 p. (Southern Uplands)

——1876b, Llandovery rocks in the Lake District: Geol. Mag., dec. 2, v. 3, p. 477-480. (Northern England)

——1878a, Recent discoveries among the Silurians of south Scotland: Glasgow Geol. Soc. Trans., v. 6, p. 78-84 and 132. (Southern Uplands)

——1878b, The Moffat Series: Geol. Soc. London Quart. Jour., v. 34, p. 240-346. (Southern Uplands)

——1879, On the tripartite classification of the lower Palaeozoic rocks: Geol. Mag., dec. 2, v. 6, p. 1-15. (Correlation—General)

——1882, The Girvan succession: Geol. Soc. London Quart. Jour., v. 38, p. 537-666. (Midland Valley)

——1886, Geology and physiology (Birmingham district), *in* Handbook of Birmingham: British Assoc. Adv. Sci. Handbook, p. 213-265 and 353-357. (Northern Welsh Borderland)

——1887, On the Palaeozoic rocks of the Birmingham district: Rept. British Assoc. Adv. Sci., Birmingham, 1886, Trans. Secs., p. 621-622. (Northern Welsh Borderland)

——1891, The geology of the Dudley district: Midland Naturalist, v. 14, p. 269-273. (Northern Welsh Borderland)

——1898, A sketch of the geology of the Birmingham district: Geol. Assoc. London Proc., v. 15, p. 313-416. (Northern Welsh Borderland)

——1899, The survey memoir on the Southern Uplands: A review: Geol. Mag., dec. 4, v. 6, p. 472-479 and 510-520. (Southern Uplands)

——1913, The Birmingham country, its geology and physiography (reprint from British Assoc. Adv. Sci. Handbook): Birmingham, Cornish Brothers, p. 1-53. (Northern Welsh Borderland)

Lapworth, C., and Watts, W. W., 1894, The geology of south Shropshire: Geol. Assoc. London Proc., v. 13, p. 297-355. (Northern Welsh Borderland)

——1910, Shropshire—Geology in the field. Pt. 4 (Jubilee volume): Geol. Assoc. London, p. 759-769. (Northern Welsh Borderland)

Lapworth, C., and Wilson, J., 1871, On the Silurian rocks of the counties of Roxburgh and Selkirk: Geol. Mag., dec. 1, v. 8, p. 456-464. (Southern Uplands)

——1872, On the Silurian rocks of the counties of Roxburgh and Selkirk: Rept. British Assoc. Adv. Sci., Edinburgh, 1871, Trans. Secs., p. 103-104. (Southern Uplands)

Lapworth, C., Watts, W. W., and Harrison, W. J., 1897-1898, Sketch of the geology of the Birmingham district: Geol. Assoc. London Proc., v. 18, p. 313-389. (Northern Welsh Borderland)

Lapworth, H., 1900, The Silurian sequence of Rhayader: Geol. Soc. London Quart. Jour., v. 56, p. 67-137. (Central Wales)

——1906, The geology of central Wales: Geol. Assoc. London Proc., v. 19, p. 160-172. (Central Wales)

La Touche, J. D., 1884, A handbook to the geology of Shropshire: London, Edward Stanford, 91 p. (Northern Welsh Borderland)

Lawson, J. D., 1954, The Silurian succession at Gorsley (Herefordshire): Geol. Mag., v. 91, p. 227-237. (Southern Welsh Borderland)

——1955, The geology of the May Hill inlier: Geol. Soc. London Quart. Jour., v. 111, p. 85-116. (Southern Welsh Borderland)

——1960, The succession of shelly faunas in the British Ludlovian: Internat. Geol. Cong., 21st, Copenhagen 1960, Rept., pt. 7, p. 114-125. (Wales and Welsh Borderland—General)

——1966, The Ludlovian rocks of the Welsh Borderland: Rept. British Assoc. Adv. Sci., v. 12, p. 563-570. (Wales and Welsh Borderland—General)

——1971, Some problems and principles in the classification of the Silurian System, in Babin, C., ed., Colloque ordovicien-silurien: Bur. Recherches Géol. Minières, Mèm., no. 75, p. 301-308. (Correlation—General)

——1973a, Facies and faunal changes in the Ludlovian rocks of Aymestry, Herefordshire: Geol. Jour., v. 8, p. 247-278. (Northern Welsh Borderland)

——1973b, New exposures on forestry roads near Ludlow: Geol. Jour., p. 279-284. (Northern Welsh Borderland)

Lawson, J. D., and Whitaker, J.H.McD., 1969, Correlation of the Leintwardine Beds: Geol. Jour., v. 6, p. 329-332. (Northern Welsh Borderland)

Lewis, H. P., 1934, The occurrence of fossiliferous pebbles of Salopian age in the Peel Sandstone (Isle of Man): Great Britain Geol. Survey Summary Progress for 1933, pt. II, p. 91-108. (North Wales)

——1946, Bedding-faults and related minor structures in the upper Valentian rocks near Aberystwyth: Geol. Mag., v. 83, p. 153-161. (Central Wales)

Lightbody, R., 1863, Notice of a section at Mocktree: Geol. Soc. London Quart. Jour., v. 19, p. 368-371. (Northern Welsh Borderland)

——1869, Notes on the geology of Ludlow: Geol. Mag., dec. 1, v. 6, p. 353-355. (Northern Welsh Borderland)

Lister, T. R., 1970, The acritarchs and chitinozoa from the Wenlock and Ludlow Series of the Ludlow and Millichope areas, Shropshire. Pt. 1: Palaeontographical Soc. Mon., 100 p. (Northern Welsh Borderland)

Lister, T. R., Cocks, L.R.M., and Rushton, A.W.A., 1970, The basement beds in the Bobbing borehole, Kent: Geol. Mag., v. 106, p. 601-603. (Southeast England)

Llewellyn, P. G., 1960, The Middle and Upper Silurian rocks between Longsleddale and the Shap Granite, Westmorland [Ph.D. thesis]: Cambridge, Cambridge Univ. (Northern England)

——1968, Silurian successions in northwest England: A bibliography: Soc. Bibliography Nat. History Jour., v. 5, p. 41-56. (Northern England)

Lovell, J.P.B., 1970, The palaeogeographical significance of lateral variations in the ratio of sandstone to shale and other features of the Aberystwyth Grits: Geol. Mag., v. 107, p. 147-158. (Central Wales)

Lumsden, G. I., Tullock, W., Howells, M. F., and Davies, A., 1967, The geology of the neighbourhood of Langholm: Great Britain Geol. Survey Mem., sheet H, new ser., 255 p. (Southern Uplands)

Macconochie, A., 1884, Review of the southern Silurian question: Glasgow Geol. Soc. Trans., v. 7, p. 370-372. (Southern Uplands)

MacGregor, M., 1927, The geology of the district around Edinburgh: Carstairs and Tinto. I. Carstairs district: Geol. Assoc. London Proc., p. 495-499. (Midland Valley)

MacGregor, M., and Read, H. H., 1925, Strathaven, Tinto, and Douglas districts: Great Britain Geol. Survey Summary Progress for 1924, p. 96-107. (Midland Valley)

MacKenzie, D. H., 1956, A structural profile south of Eyemouth, Berwickshire: Edinburgh Geol. Soc. Trans., v. 16, p. 248-253. (Southern Uplands)

Mackie, W., 1929, Preliminary report on the heavy minerals of the Silurian rocks of south Scotland: Rept. British Assoc. Adv. Sci., Glasgow, 1928, Trans. Secs., p. 556. (Southern Uplands)

Marr, J. E., 1878, On some well-defined life-zones in the lower part of the Silurian (Sedgwick) of the Lake District: Geol. Soc. London Quart. Jour., v. 34, p. 871-885. (Northern England)

——1880, On the Cambrian (Sedgwick) and Silurian rocks of the Dee Valley, as compared with those of the Lake District: Geol. Soc. London Quart. Jour., v. 36, p. 277-284. (North Wales)

——1881a, On some sections in the lower Palaeozoic rocks of the Craven district: Yorkshire Geol. Soc. Proc., v. 7, p. 387-399. (Northern England)

——1881b, The classification of the Cambrian and Silurian rocks: Geol. Mag., dec. 2, v. 8, p. 245-250. (Correlation—General)

——1883, The classification of the Cambrian and Silurian rocks: Cambridge, Deighton, Bell and Co., 147 p. (Correlation—General)

——1887, The lower Palaeozoic rocks near Settle: Geol. Mag., dec. 3, v. 4, p. 35-38. (Northern England)

——1892, On the Wenlock and Ludlow strata of the Lake District: Geol. Mag., dec. 3, v. 9, p. 534-541. (Northern England)

——1907, The geology of the Appleby district: Geol. Assoc. London Proc., v. 20, p. 129-149, 193-200. (Northern England)

——1910, The Lake District and neighbourhood—Lower Palaeozoic times (Jubilee volume): Geol. Assoc. London, p. 624-641. (Northern England)

——1913, The lower Palaeozoic rocks of the Cautley district (Yorkshire): Geol. Soc. London Quart. Jour., v. 69, p. 1-18. (Northern England)

——1916, The geology of the Lake District: Cambridge, Cambridge Univ. Press., 220 p. (Northern England)

——1925, Conditions of deposition of the Stockdale Shales of the Lake District: Geol. Soc. London Quart. Jour., v. 81, p. 113-136. (Northern England)

——1927, The deposition of the later Silurian rocks of the Lake District: Geol. Mag., v. 64, p. 494-500. (Northern England)

——1928, A possible chronometric scale for the graptolite-bearing strata: Palaeobiologica, v. 1, p. 161-162. (Correlation—General)

Marr, J. E., and Nicholson, H. A., 1888, The Stockdale Shales: Geol. Soc. London Quart. Jour., v. 44, p. 654-732. (Northern England)

Marr, J. E. and Roberts, T., 1885, The lower Palaeozoic rocks of the neighborhood of Haverfordwest: Geol. Soc. London Quart. Jour., v. 41, p. 476-491. (Southwest Wales)

Marshall, J. G., 1840, Description of a section across the Silurian rocks in Westmorland from Shap Granite to Casterton Fell: Rept. British Assoc. Adv. Sci., Birmingham, 1839, Trans. Secs., p. 67. (Northern England)

Marston, A., 1870, On the transition beds between the Devonian and Silurian rocks: Geol. Mag., dec. 1, v. 7, p. 408-410. (Wales and Welsh Borderland—General)

Martinsson, A., 1967, The succession and correlation of ostracode faunas in the Silurian of Gotland: Geol. Fören. Stockholm Förh., v. 89, p. 350-386. (Correlation—General)

McCabe, P. J., 1972, The Wenlock and Ludlow strata of the Austwick and Horton-in-Ribblesdale inlier of north-west Yorkshire: Yorkshire Geol. Soc. Proc., v. 39, p. 167-174. (Northern England)

McCoy, F., 1846, A synopsis of the Silurian fossils of Ireland: Dublin, University Press, 72 p. (Ireland)

McHenry, A., 1912, Report on the Dingle Bed rocks: Royal Irish Acad. Proc., v. 29, sec. B, p. 229-234. (Ireland)

McKerrow, W. S., and Campbell, C. J., 1960, The stratigraphy and structure of the lower Palaeozoic rocks of north-west Galway: Royal Dublin Soc. Sci. Proc., ser. A., v. 1, p. 27-51. (Ireland)

McKerrow, W. S., and Ziegler, A. M., 1972a, Silurian paleogeographic development of the proto-Atlantic Ocean: Internat. Geol. Cong., 24th, Montreal 1972, Rept., pt. 6: Gardenvale, Quebec, Harpell's Press, p. 4-10. (British Isles—General)

——1972b, Palaeozoic oceans: Nature Phys. Sci., v. 240, p. 92-94. (British Isles—General)

McLaren, D. J., and Miller, T. G., 1948, Notes on the geology of Killary Harbour: Geol. Mag., v. 85, p. 217-221. (Ireland)

Mitchell, G. H., 1956, The geological history of the Lake District: Yorkshire Geol. Soc. Proc., v. 30, p. 407-463. (Northern England)

Mitchell, G. H., and Mykura, W., 1962, The geology of the neighbourhood of Edinburgh: Great Britain Geol. Survey Mem., 145 p. (Midland Valley)

Morgan, C. L., and Reynolds, S. H., 1901, The igneous rocks and associated sedimentary beds of the Tortworth inlier: Geol. Soc. London Quart. Jour., v. 57, p. 267-284. (Southern Welsh Borderland)

Morgan, J. B., 1891, On the strata forming the base of the Silurian in north-east Montgomery-shire: Rept. British Assoc. Adv. Sci., Leeds, 1890, Trans. Secs., p. 816. (Central Wales)

Murchison, R. I., 1833, On the sedimentary deposits which occupy the western parts of Shropshire and Herefordshire and are prolonged from N.E. to S.W., through Radnorshire, Brecknockshire, and Carmarthenshire, with descriptions of the accompanying rocks of intrusive or igneous characters: Geol. Soc. London Proc., v. 1, p. 474-477. (Wales and Welsh Borderland—General)

——1834a, On the Upper Greywacke Series of England and Wales: Edinburgh New Philos. Jour., v. 17, p. 365-368. (Wales and Welsh Borderland—General)

——1834b, On the structure and classification of the Transition Rocks of Shropshire, Herefordshire, and parts of Wales, and on the lines of disturbance which have affected that series of deposits, including the valley of elevation of Woolhope: Geol. Soc. London Proc., v. 2, p. 13-18. (Wales and Welsh Borderland—General)

——1835, On the Silurian system of rocks: Philos. Mag., v. 7, p. 46-52. (Correlation—General)

——1836a, On the geological structure of Pembrokeshire, more particularly on the extension of the Silurian system of rocks into the coast cliffs of that country: Geol. Soc. London Proc., v. 2, p. 226-230. (Southwest Wales)

——1836b, On the Silurian and other rocks of the Dudley and Wolverhampton coalfield followed by a sketch proving the Lickey Quartz rock to be of the same age as the Caradoc Sandstone: Geol. Soc. London Proc., v. 2, p. 407-414. (Northern Welsh Borderland)

——1839, The Silurian system (2 vols.): London, John Murray, 768 p. (British Isles—General)

——1847, On the meaning originally attached to the term, "Cambrian System," and on the evidence since obtained of its being geologically synonymous with the previously established term, "Lower Silurian": Geol. Soc. London Quart. Jour., v. 3, p. 165-179. (Correlation—General)

——1851, On the Silurian rocks of the south of Scotland: Geol. Soc. London Quart. Jour., v. 7, p. 139-169. (Scotland—General)

——1852a, The Silurian System: Edinburgh New Philos. Jour., v. 52, p. 355-358. (Correlation—General)

——1852b, On the meaning of the term "Silurian System" as adopted by geologists in various countries during the last ten years: Geol. Soc. London Quart. Jour., v. 8, p. 173-184. (Correlation—General)

——1853, On some of the remains in the bone-bed of the upper Ludlow: Geol. Soc. London Quart. Jour., v. 9, p. 16-17. (Northern Welsh Borderland)

——1854, Siluria: London, John Murray, 523 p. (British Isles—General)

——1856, On the discovery, by Robert Slimon, of fossils in the uppermost Silurian rocks near Lesmahago in Scotland, with observations on the relations of the Palaeozoic strata in that part of Lanarkshire: Geol. Soc. London Quart. Jour., v. 12, p. 15-24. (Midland Valley)

——1857, On the bone-beds of the upper Ludlow rock and base of the Old Red Sandstone: Rept. British Assoc. Adv. Sci., Cheltenham, 1856, Trans. Secs., p. 70-71. (Northern Welsh Borderland)

——1859, Siluria (3d ed.): London, John Murray, 566 p. (British Isles—General)

——1867, Siluria (4th ed.): London, John Murray, 566 p. (British Isles—General)

——1872, Siluria (5th ed.): London, John Murray, 566 p. (British Isles—General)

Mykura, W., 1951, The age of the Malvern folding: Geol. Mag., v. 88, p. 386-392. (Southern Welsh Borderland)

Newall, G., 1966, A faunal and sedimentary study of the Aymestry limestones and adjacent beds in parts of Herefordshire and Shropshire [Ph.D. thesis]: Manchester, Manchester Univ. (Northern Welsh Borderland)

Nicholson, H. A., 1868, Essay on the geology of Cumberland and Westmorland: London, Robert Hardwicke, 93 p. (Northern England)

——1870, On the correlation of the Silurian deposits of the north of England with those of the south of Scotland: Edinburgh Geol. Soc. Trans., v. 2, p. 105-113. (Northern England)

——1872, On the Silurian rocks of the English Lake District: Geol. Assoc. London Proc., v. 3, p. 105-114. (Northern England)

Nicholson, H. A., and Lapworth, C., 1876, On the central group of the Silurian series of the north of England: Rept. British Assoc. Adv. Sci., Bristol, 1875, Trans. Secs., p. 78-79. (Northern England)

Nicholson, H. A., and Marr, J. E., 1891, Cross Fell inlier: Geol. Soc. London Quart. Jour., v. 47, p. 500-529. (Northern England)

Nicol, J., 1848, On the geology of the Silurian rocks in the valley of the Tweed: Geol. Soc. London Quart. Jour., v. 4, p. 195-209. (Southern Uplands)

——1850, Observations on the Silurian strata of the southeast of Scotland: Geol. Soc. London Quart. Jour., v. 6, p. 53-65. (Southern Uplands)

Norman, T. N., 1961, The geology of the Silurian strata in the Blawith area, Furness [Ph.D. thesis]: Birmingham, Birmingham Univ. (Northern England)

North, F. J., 1915, Note on the Silurian inlier near Cardiff: Geol. Mag., dec. 6, v. 2, p. 385-387. (Southern Welsh Borderland)

O'Connor, B., 1969, The structure and sedimentation in part of the Ordovician and Silurian in the Tarn Hows area of the central-southern Lake District [M.Sc. thesis]: Liverpool, Liverpool Univ. (Northern England)

Okada, H., 1966, Non-greywacke "turbidite" sandstones in the Welsh geosyncline: Sedimentology, v. 7, p. 211-232. (Central Wales)

——1967a, Composition and cementation of some lower Palaeozoic grits in Wales: Kyushu Univ. Fac. Sci. Mem., ser. D, Geology, v. 18, p. 261-276. (Wales and Welsh Borderland— General)

——1967b, Some problems of geosynclinal clastic sediments: Kagaku (Science), v. 37, p. 270-276. (Wales and Welsh Borderland—General)

Otley, J., 1820, Remarks on the succession of rocks in the district of the Lakes: Philos. Mag., v. 56, p. 257-261. (Northern England)

Owen, T. R., 1967, From the south: A discussion: Geol. Assoc. London Proc., v. 78, p. 595-601. (Wales and Welsh Borderland—General)

Palmer, D., 1970a, A stratigraphical synopsis of the Long Mountain, Montgomeryshire and Shropshire: Geol. Soc. London Proc., no. 1660, p. 341-346. (Northern Welsh Borderland)

——1970b, *Monograptus ludensis* zone graptolites from the Devilsbit Mountain district, Tipperary: Royal Dublin Soc. Sci. Proc., ser. A, v. 3, p. 335-342. (Ireland)

——1971, The Ludlow graptolites *Neodiversograptus nilsonni* and *Cucullograptus* (*lobograptus*) *progenitor*: Lethaia, v. 4, p. 357-384. (Correlation—General)

——1972, The geology of the Longmountain, Montgomeryshire and Shropshire [Ph.D. thesis]: Dublin, Trinity College, Dublin Univ. (Northern Welsh Borderland)

Peach, B. N., and Horne, J., 1899, The Silurian rocks of Britain. Vol. 1: Scotland: Great Britain Geol. Survey Mem., 749 p. (Scotland—General)

Peach, B. N., Clough, C. T., Hinxman, L. W., Grant Wilson, J. S., Crampton, C. B., Maufe, H. B., and Bailey, E. B., 1910, The geology of the neighbourhood of Edinburgh (2d ed.): Great Britain Geol. Survey Mem., sheet 32, 445 p. (Midland Valley)

Penn, J.S.W., 1971, Bioherms in the Wenlock Limestones of the Malvern area (Herefordshire, England), *in* Babin, C., ed., Colloque, ordovicien-silurien: Bur. Recherches Géol. et Minières, Mém., no. 73, p. 129-138. (Southern Welsh Borderland)

Penn, J.S.W., French, J., Whitten, D.G.A., and Vinnicombe, J., 1971, The Malvern Hills: Geol. Assoc. London, Guide no. 4, 36 p. (Southern Welsh Borderland)

Phillips, J., 1842, On the occurrence of shells and corals in a conglomerate bed, adherent to the face of the trap rocks of the Malvern Hills, and full of rounded and angular fragments of those rocks: Philos. Mag., ser. 3, v. 21, p. 288-293. (Southern Welsh Borderland)

——1848, The Malvern Hills compared with the Palaeozoic districts of Abberley Woolhope, May Hill, Tortworth, and Usk: Great Britain Geol. Survey Mem., v. 2, pt. 1, 386 p. (Southern Welsh Borderland)

Phillips, W.E.A., 1974, The stratigraphy, sedimentary environments, and palaeogeography of the Silurian strata of Clare Island, County Mayo, Ireland: Geol. Soc. London Jour., v. 130, p. 19-41. (Ireland)

Phillips, W.E.A., and Skevington, D., 1968, The lower Palaeozoic rocks of the Lough Acanon area, County Cavan, Ireland: Royal Dublin Soc. Sci. Proc., v. 3, ser. A, p. 141-148. (Ireland)

Phillips, W.E.A., Rickards, R. B., and Dewey, J. F., 1970, The lower Palaeozoic rocks of the Louisburgh Area, County Mayo: Royal Irish Acad. Proc., v. 70, sec. B, p. 195-210. (Ireland)

Phipps, C. B., 1962, The revised Ludlovian stratigraphy of the type area—A discussion:

Geol. Mag., v. 99, p. 385-392. (Correlation—General)

Phipps, C. B., and Reeve, F.A.E., 1964, The Pre-Cambrian-Palaeozoic boundary of the Malverns: Geol. Mag., v. 101, p. 397-408. (Southern Welsh Borderland)

——1967, Stratigraphy and geological history of the Malvern, Abberley, and Ledbury Hills: Geol. Jour., v. 5, p. 339-368. (Southern Welsh Borderland)

——1969, Structural geology of the Malvern, Abberley, and Ledbury Hills: Geol. Soc. London Quart. Jour., v. 125, p. 1-38. (Southern Welsh Borderland)

Piper, D.J.W., 1967, A new interpretation of the Llandovery sequence of North Connemara, Eire: Geol. Mag., v. 104, p. 253-267. (Ireland)

——1969, Geosyncline-margin sedimentary rocks in Silurian of West Connacht, Ireland, in Kay, M., ed., North Atlantic—Geology and continental drift: Am. Assoc. Petroleum Geologists Mem. 12, p. 289-297. (Ireland)

——1970, A Silurian deep sea fan deposit in western Ireland and its bearing on the nature of turbidity currents: Jour. Geology, v. 78, p. 509-522. (Ireland)

——1972a, Sedimentary environments and palaeogeography of the late Llandovery and earliest Wenlock of North Connemara, Ireland: Geol. Soc. London Jour., v. 128, p. 33-52. (Ireland)

——1972b, Trench sedimentation in the lower Palaeozoic of the Southern Uplands: Scottish Jour. Geology, v. 8, p. 289-291. (Southern Uplands)

Pocock, R. W., 1930, The *Petalocrinus* limestone horizon at Woolhope (Herefordshire): Geol. Soc. London Quart. Jour., v. 86, p. 50-63. (Southern Welsh Borderland)

Pocock, R. W., Whitehead, T. H., Wedd, C. B., and Robertson, T., 1938, Shrewsbury district, including the Hanwood Coalfield: Great Britain Geol. Survey Mem., sheet 152, new ser., 297 p. (Northern Welsh Borderland)

Pocock, R. W., Brammall, A., and Croft, W. N., 1940, Easter Field meeting, Hereford: Geol. Assoc. London Proc., v. 51, p. 52-62. (Southern Welsh Borderland)

Pollock, J., and Wilson, H. E., 1961, A new fossiliferous locality in County Down: Irish Naturalists' Jour., v. 13, p. 244-248. (Ireland)

Portlock, J. E., 1843, Report on the geology of the county of Londonderry and parts of Tyrone and Femanagh: Dublin, Her Majesty's Stationery Office, 784 p. (Ireland)

Postlethwaite, J., 1897, The geology of the English Lake District with notes on the minerals: Keswick, T. Bakewell, 78 p. (Northern England)

——1903, The geology of the English Lake District: Federated Inst. Mining Engineers Trans., v. 25, p. 302-330. (Northern England)

Potter, J. F., 1965, Rotational strike-slip faults, Llandeilo, Wales: Geol. Mag., v. 102, p. 496-500. (Central Wales)

——1967, Deformed micaceous deposits in the Downtonian of the Llandeilo region, South Wales: Geol. Assoc. London Proc., v. 78, p. 277-288. (Central Wales)

——1968, The Silurian-Devonian boundary: Geol. Mag., v. 105, p. 187-188. (Correlation—General)

Potter, J. F., and Price, J. H., 1965, Comparative sections through rocks of Ludlovian-Downtonian age in the Llandovery and Llandeilo districts: Geol. Assoc. London Proc., v. 76, p. 379-402. (Central Wales)

——1967, Written discussion to the summer (1964) field meeting in South Wales report: Geol. Assoc. London Proc., v. 77, p. 384. (Central Wales)

Prendergast, B. M., 1972, The Silurian and Devonian inlier of Knockshigowna Hill, County Tipperary, Ireland: Royal Dublin Soc. Sci. Proc., ser. A, v. 4, p. 201-211. (Ireland)

Prestwich, J., 1858, On the boring through the chalk at Harwich: Geol. Soc. London Quart. Jour., v. 14, p. 249-252. (Southeast England)

Price, N. J., 1962, The tectonics of the Aberystwyth grits: Geol. Mag., v. 99, p. 542-557. (Central Wales)

Pringle, J., 1948, British regional geology: The south of Scotland (2d ed.): Edinburgh, Her Majesty's Stationery Office, 87 p. (Southern Uplands)

Pugh, W. J., 1923, The geology of the district around Corris and Aberllefeni, Merionethshire: Geol. Soc. London Quart. Jour., v. 79, p. 508-545. (North Wales)

——1928, The geology of the district around Dinas Mawddwy, Merioneth: Geol. Soc. London Quart. Jour., v. 84, p. 345-381. (North Wales)

——1929, The geology of the district between Llangmawddwy and Llanawchllyn, Merioneth: Geol. Soc. London Quart. Jour., v. 85, p. 242-306. (North Wales)

——1949, Recent work on the lower Palaeozoic rocks: Rept. British Assoc. Adv. Sci., v. 6, p. 203-211. (British Isles—General)

Ramsay, A. C., 1853, On the physical structure and succession of some of the Lower Palaeozoic rocks of North Wales and part of Shropshire: Geol. Soc. London Quart. Jour., v. 9, p. 161-179. (North Wales)

——1863, Breaks in succession of the British Palaeozoic strata: Geol. Soc. London Quart. Jour., v. 19, p. 36-52. (British Isles—General)

——1894, The physical geology and geography of Great Britain: London, Edward Stanford, 421 p. (British Isles—General)

Ramsay, A. C., and Aveline, W. T., 1848, Sketch of the structure of parts of North and South Wales: Geol. Soc. London Quart. Jour., v. 4, p. 294-299. (Wales and Welsh Borderland—General)

Ramsay, A. C., and Salter, J. W., 1866, The geology of North Wales and an appendix on the fossils, with plates: Great Britain Geol. Survey Mem., v. 3, 381 p. (North Wales)

——1881, The geology of North Wales with an appendix on the fossils (2d ed.): Great Britain Geol. Survey Mem., v. 3, 611 p. (North Wales)

Rastall, R. H., 1925, Petrographical notes on the Stockdale shales: Geol. Soc. London Quart. Jour., v. 81, p. 131-133. (Northern England)

Raw, Frank, 1952, Structure and origin of the Malvern Hills: Geol. Assoc. London Proc., v. 63, p. 227-239. (Southern Welsh Borderland)

Raw, Frank, Shotton, F. W., and Wills, L. J., 1935, Whitsun field meeting, 1935. The Birmingham district: Geol. Assoc. London Proc., v. 46, p. 391-398. (Northern Welsh Borderland)

Rayner, D. H., 1967, The stratigraphy of the British Isles: Cambridge, Cambridge Univ. Press, 453 p. (British Isles—General)

Read, H. H., 1927, The geology of the district around Edinburgh: Carstairs and Tinto. II. The Tinto district: Geol. Assoc. London Proc., v. 38, p. 299-342. (Midland Valley)

Reading, H. G., and Poole, A. B., 1961, A Llandovery shoreline from the southern Malverns: Geol. Mag., v. 98, p. 295-300. (Southern Welsh Borderland)

——1962, Malvern structures: Geol. Mag., v. 99, p. 377-379. (Southern Welsh Borderland)

Reed, F.R.C., 1907, The base of the Silurian near Haverfordwest: Geol. Mag., dec. 5, v. 4, p. 535-537. (Southwest Wales)

——1917, The Ordovician and Silurian Brachiopoda of the Girvan district: Royal Soc. Edinburgh Trans., v. 51, p. 795-998. (Midland Valley)

——1934, Downtonian fossils from the Anglo-Welsh area: Geol. Soc. London Quart. Jour., v. 90, p. 571-584. (Wales and Welsh Borderland—General)

Reed, F.R.C., and Reynolds, S. H., 1908a, On the fossiliferous Silurian rocks of the southern half of the Tortworth inlier: Geol. Soc. London Quart. Jour., v. 64, p. 512-545. (Southern Welsh Borderland)

——1908b, Silurian fossils from certain localities in the Tortworth inlier: Bristol Naturalists' Soc. Proc., v. 2, p. 32-40. (Southern Welsh Borderland)

Reid, C., 1907, The geology of the country around Mevagissey: Geol. Survey Great Britain Mem., sheet 353, 73 p. (Southwest England)

Reynolds, S. H., 1907, A Silurian inlier in the eastern Mendips: Geol. Soc. London Quart. Jour., v. 63, p. 217-240. (Southern Welsh Borderland)

——1910, The Palaeozoic rocks of Gloucestershire and Somerset (Jubilee volume): Geol. Assoc. London, p. 308-328. (Southern Welsh Borderland)

——1912, Further work on the Silurian rocks of the eastern Mendips: Bristol Naturalists' Soc. Proc., v. 3, p. 76-82. (Southern Welsh Borderland)

——1924, The igneous rocks of the Tortworth inlier: Geol. Soc. London Quart. Jour., v. 80, p. 106-112. (Southern Welsh Borderland)

——1929, The geology of the Bristol district: Geol. Assoc. London Proc., v. 40, p. 77-103. (Southern Welsh Borderland)

——1946, The Aust section: Cotteswold Naturalists' Field Club Proc., v. 29, p. 29-39. (Southern Welsh Borderland)

Reynolds, S. H., Woodward, H. B., Morgan, C. L., and Winwood, H. H., 1909, Investigation of the pre-Devonian rocks of the Mendips and the Bristol area: Rept. British Assoc. Adv. Sci., Dublin, 1908, Trans. Secs., p. 286-291. (Southern Welsh Borderland)

Rhodes, F.H.T., 1953, Some British Lower Palaeozoic conodont faunas: Royal Soc. London Philos. Trans., ser. B, v. 237, p. 261-334. (Correlation—General)

Rhodes, F.H.T., and Newall, G., 1963, Occurrence of *Kockelella variabilis* Wallister in the Aymestry Limestone of Shropshire: Nature, v. 199, p. 166-167. (Northern Welsh Borderland)

Richardson, J. B., and Lister, T. R., 1969, Upper Silurian and Lower Devonian spore assemblages from the Welsh Borderland and South Wales: Palaeontology, v. 12, p. 201-252. (Wales and Welsh Borderland—General)

Richardson, Linsdall, 1929, The country around Moreton in Marsh: Great Britain Geol. Survey Mem., sheet 217, new ser., 162 p. (Southeast England)

Rickards, R. B., 1964a, Further discussion on "The stratigraphy and structure of the Silurian (Wenlock) rocks south-east of Hawick, Roxburghshire, Scotland," by P. T. Warren: Geol. Soc. London Quart. Jour., v. 120, p. 548-549. (Southern Uplands)

——1964b, The graptolitic mudstone and associated facies in the Silurian strata of the Howgill Fells: Geol. Mag., v. 101, p. 435-451. (Northern England)

——1965, New Silurian graptolites from the Howgill Fells: Palaeontology, v. 8, p. 247-271. (Northern England)

——1967, The Wenlock and Ludlow succession in the Howgill Fells (north-west Yorkshire and Westmorland): Geol. Soc. London Quart. Jour., v. 123, p. 215-251. (Northern England)

——1969, Wenlock graptolite zones in the English Lake District: Geol. Soc. London Proc., no. 1654, p. 61-65. (Northern England)

——1970a, The Llandovery (Silurian) graptolites of the Howgill Fells, northern England: Palaeontographical Soc. Mon., v. 123, 108 p. (Northern England)

——1970b, Age of the Middle Coldwell Beds: Geol. Soc. London Proc., no. 1663, p. 111-114. (Northern England)

——1973, On some highest Llandovery red beds and graptolite assemblages in Britain and Eire: Geol. Mag., v. 110, p. 70-72. (British Isles—General)

Rickards, R. B., and Archer, J. B., 1969, The lower Palaeozoic rocks near Tomgraney, County Clare: Royal Dublin Soc. Sci. Proc., ser. A, v. 3, p. 219-229. (Ireland)

Rickards, R. B., and Smyth, W. R., 1968, The Silurian graptolites of Mayo and Galway: Royal Dublin Soc. Sci. Proc., ser. A, v. 3, p. 129-135. (Ireland)

Rickards, R. B., Burns, V., and Archer, J. B., 1973, The Silurian sequence at Balbriggan, County Dublin: Royal Irish Acad. Proc., v. 73, sec. B, p. 303-316. (Ireland)

Ritchie, A., 1963, Palaeontological studies on Scottish Silurian fish beds [Ph.D. thesis]: Edinburgh, Edinburgh Univ. (Midland Valley)

Ritchie, A., 1967, *Ateleaspis tessellata* Traquair, a non-cornuate cephalaspid from the Upper Silurian of Scotland: Linnean Soc. London Jour., Zoology, v. 47, p. 69-81. (Midland Valley)

Ritchie, M., and Eckford, R.J.A., 1936, The Haggis Rock of the Southern Uplands: Edinburgh Geol. Soc. Trans., v. 13, p. 371-377. (Southern Uplands)

Roberts, G. E., and Randall, J., 1863, On the Upper Silurian passage-beds at Linley, Salop: Geol. Soc. London Quart. Jour., v. 19, p. 229-232. (Northern Welsh Borderland)

Roberts, R. O., 1929, The geology of the district around Abbey-Cwmhir, Radnorshire: Geol. Soc. London Quart. Jour., v. 85, p. 651-676. (Central Wales)

Robertson, T., 1926, The section of the new railway tunnel through the Malvern Hills at Colwall: Great Britain Geol. Survey Summary Progress for 1925, p. 162-175. (Southern Welsh Borderland)

——1927a, The highest Silurian rocks of the Wenlock district: Great Britain Geol. Survey Summary Progress for 1926, p. 80-97. (Northern Welsh Borderland)

——1927b, The geology of the South Wales coalfield. Pt. II. The country around Abergavenny (2d ed.): Great Britain Geol. Survey Mem., sheet 232, new ser., 145 p. (Southern Welsh Borderland)

——1928, The Siluro-Devonian junction in England: Geol. Mag., v. 65, p. 385-400. (Correlation—General)

Rolfe, W.D.I., 1960, The Silurian inlier of Carmichael, Lanarkshire: Royal Soc. Edinburgh Trans., v. 64, p. 245-260. (Midland Valley)

——1961, The geology of the Hagshaw Hills Silurian inlier, Lanarkshire: Geol. Soc. London Proc., no. 1585, p. 48-52. (Midland Valley)

——1962, The geology of the Hagshaw Hills Silurian inlier, Lanarkshire: Edinburgh Geol. Soc. Trans., v. 18, p. 240-269. (Midland Valley)

Rolfe, W.D.I., and Fritz, M. A., 1966, Recent evidence for the age of the Hagshaw Hills Silurian inlier, Lanarkshire: Scottish Jour. Geology, v. 2, p. 151-164. (Midland Valley)

Ross, G., Tait, D., and Pringle, J., 1930, On the age of the lowest Silurian rocks of the Hagshaw Hills and Lesmahagow inliers [abs.]: Glasgow Geol. Soc. Trans., v. 18, p. 634. (Midland Valley)

Ruddy, T., 1879, On the upper part of the Cambrian (Sedgwick) and base of the Silurian in North Wales: Geol. Soc. London Quart. Jour., v. 35, p. 200-208. (North Wales)

Rust, B. R., 1965a, The stratigraphy and structure of the Whithorn area of Wigtownshire, Scotland: Scottish Jour. Geology, v. 1, p. 101-133. (Midland Valley)

——1965b, The sedimentology and diagenesis of Silurian turbidites in south-east Wigtownshire, Scotland: Scottish Jour. Geology, v. 1, p. 231-246. (Midland Valley)

Rutley, F., 1887, On the rocks of the Malvern Hills: Geol. Soc. London Quart. Jour., v. 43, p. 481-516. (Southern Welsh Borderland)

St. Joseph, J.K.S., 1935, A critical examination of *Stricklandia* (=*Stricklandinia*) *lirata* (J. de C. Sowerby) 1839 *forma typica*: Geol. Mag., v. 72, p. 401-422. (Correlation—General)

——1938, The Pentameracea of the Oslo region: Norsk Geol. Tidsskr., v. 17, p. 225-336. (Correlation—General)

Salter, J. W., 1867, On the May Hill Sandstone: Geol. Mag., dec. 1, v. 4, p. 201-205. (Southern Welsh Borderland)

——1869, Some new points in the geology of the Usk district: Woolhope Naturalists' Field Club Trans. 1868, p. 174. (Southern Welsh Borderland)

Salter, J. W., and Aveline, W. T., 1854, On the Caradoc Sandstone of Shropshire: Geol. Soc. London Quart. Jour., v. 10, p. 62-75. (Northern Welsh Borderland)

Sanzen-Baker, I., 1972, Stratigraphical relationships and sedimentary environments of the Silurian-early Old Red Sandstone of Pembrokeshire: Geol. Assoc. London Proc., v.

83, p. 139-164. (Southwest Wales)

Scoffin, T. P., 1971, The conditions of growth of the Wenlock reefs of Shropshire (England): Sedimentology, v. 17, p. 173-219. (Northern Welsh Borderland)

Sedgwick, A., 1838, A synopsis of the English series of stratified rocks inferior to the Old Red Sandstone—With an attempt to determine the successive natural groups and formations: Geol. Soc. London Proc., v. 2, p. 675-685. (Correlation—General)

——1841, Supplement to a "Synopsis of the English series of stratified rocks inferior to the Old Red Sandstone," with additional remarks on the relations of the Carboniferous Series and Old Red Sandstone of the British Isles: Geol. Soc. London Proc., v. 3, p. 545-554. (British Isles—General)

——1843a, Outline of the geological structure of North Wales: Geol. Soc. London Proc., v. 4, p. 212-224. (North Wales)

——1843b, On the older Palaeozoic (Protozoic) rocks of North Wales: Geol. Soc. London Proc., v. 4, p. 251-268. (North Wales)

——1845a, On the older Palaeozoic rocks of North Wales: Geol. Soc. London Quart. Jour., v. 1, p. 5-22. (North Wales)

——1845b, On the comparative classification of the fossiliferous strata of North Wales, with the corresponding deposits of Cumberland, Westmoreland, and Lancashire: Geol. Soc. London Quart. Jour., v. 1, p. 442-450. (Correlation—General)

——1846, On the classification of the fossiliferous slates of Cumberland, Westmoreland, and Lancashire (being a supplement to a paper read to the Society, March 12, 1845) Pt. I: Geol. Soc. London Quart. Jour., v. 2, p. 106-131. (Correlation—General)

——1847, On the classification of the fossiliferous slates of North Wales, Cumberland, Westmoreland, and Lancashire (being a supplement to a paper read to the Society, March 12, 1845) Pt. II: Geol. Soc. London Quart. Jour., v. 3, p. 133-164. (Correlation—General)

——1851-1852, A synopsis of the classification of the British Palaeozoic rocks: Fasc. 1-1851, Fasc. 2-1852, Cambridge and London, Cambridge Univ. Press. (Correlation—General)

——1852a, On the lower Palaeozoic rocks at the base of the Carboniferous chain between Ravenstonedale and Ribblesdale: Geol. Soc. London Quart. Jour., v. 8, p. 35-54. (Northern England)

——1852b, On the classification and nomenclature of the lower Palaeozoic rocks of England and Wales: Geol. Soc. London Quart. Jour., v. 8, p. 136-168. (Correlation—General)

——1853, On a proposed separation of the so-called Caradoc Sandstone into two distinct groups: (1) May Hill Sandstone, (2) Caradoc Sandstone: Geol. Soc. London Quart. Jour., v. 9, p. 215-230. (Wales and Welsh Borderland—General)

——1854a, On the May Hill Sandstone and the Palaeozoic System of the British Isles [abs.]: Geol. Soc. London Quart. Jour., v. 10, p. 366-367. (British Isles—General)

——1854b, On the classification and nomenclature of the older Palaeozoic rocks of Britain: Rept. British Assoc. Adv. Sci., Hull, 1853, Trans. Secs., p. 54-61. (Correlation—General)

——1854c, On the Mayhill Sandstone and the Palaeozoic System of England: Philos. Mag., ser. 4, v. 8, p. 301-317, 359-370, and 472-506. (British Isles—General)

——1855, A synopsis of the British Palaeozoic rocks. With a systematic description of the British Palaeozoic fossils in the geological museum of the University of Cambridge, by Professor F. McCoy: London and Cambridge, Cambridge Univ. Press, 661 p. (British Isles—General)

——1857, Description of a series of dislocations which have moved the Cambrian and Silurian rocks between Leven Sands and Duddon Sands: Philos. Mag., ser. 4, v. 16, p. 155-158. (Northern England)

Sedgwick, A., and Murchison, R. I., 1836, On the Silurian and Cambrian Systems, exhibiting the order in which the older sedimentary strata succeed each other in England and

Wales: Rept. British Assoc. Adv. Sci., Dublin, 1835, Trans. Secs., p. 59-61. (Wales and Welsh Borderland—General)

Sedgwick, A., and Murchison, R. I., 1852, The Cambrian and Silurian discussion: Edinburgh New Philos. Jour., v. 53, p. 102. (Correlation—General)

Shackleton, R. M., 1940, The succession of rocks in the Dingle peninsula, County Kerry: Royal Irish Acad. Proc., sec. B, v. 46, p. 1-12. (Ireland)

——1953, The structural evolution of North Wales: Geol. Jour., v. 1, p. 261-297. (North Wales)

Sharpe, D., 1843, On the Silurian rocks of the south of Westmoreland and north Lancashire: Geol. Soc. London Proc., v. 4, p. 23. (Northern England)

——1846, Contributions to the geology of North Wales: Geol. Soc. London Quart. Jour., v. 2, p. 283-316. (North Wales)

Shaw, R.W.L., 1969, Beyrichian ostracodes from the Downtonian of Shropshire: Geol. Fören. Stockholm Förh., v. 91, p. 52-72. (Northern Welsh Borderland)

——1971a, The faunal stratigraphy of the Kirkby Moor Flags of the type area near Kendal, Westmorland: Geol. Jour., v. 7, p. 359-380. (Northern England)

——1971b, Ostracoda from the Underbarrow, Kirkby Moor, and Scout Hill Flags (Silurian) near Kendal, Westmorland: Palaeontology, v. 14, p. 595-611. (Northern England)

Shelford, P. H., 1964, The Malvern Line: Geol. Mag., v. 101, p. 566-567. (Southern Welsh Borderland)

Shergold, J. H., and Bassett, M. G., 1970, Facies and faunas at the Wenlock/Ludlow boundary of Wenlock Edge, Shropshire: Lethaia, v. 3, p. 113-142. (Northern Welsh Borderland)

Shergold, J. H., and Shirley, J., 1968, The faunal-stratigraphy of the Ludlovian rocks between Craven Arms and Bourton near Much Wenlock, Shropshire: Geol. Jour., v. 6, p. 119-138. (Northern Welsh Borderland)

Sherlock, R. L., Casey, R., Holmes, S.C.A., and Wilson, V., 1962, British regional geology: London and Thames Valley: London, Her Majesty's Stationery Office, 62 p. (Southeast England)

Shirley, J., 1938, Some aspects of the Siluro-Devonian boundary problem: Geol. Mag., v. 75, p. 353-362. (Correlation—General)

——1939, Note on the occurrence of Dayia navicula (J. de C. Sowerby) in the lower Ludlow (Silurian) rocks of Shropshire: Geol. Mag., v. 76, p. 360-361. (Northern Welsh Borderland)

——1952, The Ludlow rocks north of Craven Arms: Geol. Assoc. London Proc., v. 63, p. 201-206. (Northern Welsh Borderland)

Shotton, F. W., 1927, The conglomerates of the Enville Series of the Warwickshire coalfield: Geol. Soc. London Quart. Jour., v. 83, p. 604-621. (Northern Welsh Borderland)

——1935, The stratigraphy and tectonics of the Cross Fell inlier: Geol. Soc. London Quart. Jour., v. 91, p. 639-704. (Northern England)

Shotton, F. W., and Trotter, F. M., 1936, Summer field meeting, 1936: The Cross Fell inlier and Stainmore: Geol. Assoc. London Proc., v. 47, p. 376-387. (Northern England)

Simpson, B., 1940, The Salopian rocks of the Clwydian Range between the Bodfari Gap and Moel Llys-y-coed, Flintshire: Geol. Assoc. London Proc., v. 51, p. 188-206. (North Wales)

——1954, Field meeting in South Wales, Swansea district: Geol. Assoc. London Proc., v. 65, p. 328-331. (Central Wales)

——1971, The Palaeozoic succession in the Black Mountains between Pontardawe and Llandilo, in Bassett, D. A., and Bassett, M. G., eds., Geological excursions in South Wales and the Forest of Dean: Cardiff, Wales Univ. Press, p. 143-154. (Central Wales)

Skevington, D., 1971, Monograptus priodon (Bronn) from the MacDuff Group (upper Dalradian) of Banffshire, Scotland: Geol. Mag., v. 108, p. 485-487. (Scotland—General)

Small, E. W., 1899, A note on some Skomer photographs: Cardiff Nat. Soc. Trans., v.

30, p. 60-61. (Southwest Wales)

Smith, A. J., and Long, G. H., 1969, The upper Llandovery sediments of Wales and the Welsh Borderlands, *in* Wood, A., ed., The Pre-Cambrian and lower Palaeozoic rocks of Wales: Cardiff, Wales Univ. Press, p. 239-253. (Wales and Welsh Borderland—General)

Smith, B., 1935, The Mynydd Cricor inlier: Geol. Assoc. London Proc., v. 46, p. 187-192. (North Wales)

Smith, E. G., Hawkins, T.R.W., Warren, P. T., and Wilson, H. E., 1965, A note on the pattern of faulting in the Ludlow rocks of northwestern Denbighshire: Great Britain Geol. Survey Bull., no. 23, p. 1-8. (North Wales)

Smith, J.D.D., 1957, Graptolites with associated sedimentary grooving: Geol. Mag., v. 94, p. 425-428. (Southeast England)

——1959, Sectional reports. 1. Palaeontological Department. A. England, Wales, and Northern Ireland: Great Britain Geol. Survey Summary Progress for 1958, p. 47. (Southeast England)

Smyth, L. B., 1939, The geology of south-east Ireland, together with parts of Limerick, Clare, and Galway: Geol. Assoc. London Proc., v. 50, p. 287-351. (Ireland)

Sollas, W. J., 1879, On the Silurian district of Rhymney and Pen-y-lan, Cardiff: Geol. Soc. London Quart. Jour., v. 35, p. 475-507. (Southern Welsh Borderland)

Squirrell, H. C., and Downing, R. A., 1969, Geology of the South Wales coalfield. Pt. I. The country around Newport (Mon.): Great Britain Geol. Survey Mem., sheet 249, new ser., 333 p. (Southern Welsh Borderland)

Squirrell, H. C., and Tucker, E. V., 1960, The geology of the Woolhope inlier, Herefordshire: Geol. Soc. London Quart. Jour., v. 116, p. 139-185. (Southern Welsh Borderland)

Stamp, L. D., 1919, The highest Silurian rocks of the Clun-forest district (Shropshire): Geol. Soc. London Quart. Jour., v. 74, p. 221-246. (Northern Welsh Borderland)

——1922, La base du Système Devonien en Angleterre: Geol. Soc. Belgique Bull., v. 31, p. 87-98. (British Isles—General)

——1923, The base of the Devonian with special reference to the Welsh Borderland: Geol. Mag., v. 60, p. 276-282, 331-336, and 367-372. (Wales and Welsh Borderland—General)

Stanton, W. I., 1960, The lower Palaeozoic rocks of south-west Murrisk, Ireland: Geol. Soc. London Quart. Jour., v. 116, p. 269-296. (Ireland)

Størmer, L., 1967, Some aspects of the Caledonian geosyncline and foreland west of the Baltic Shield: Geol. Soc. London Quart. Jour., v. 123, p. 183-214. (British Isles—General)

Storrie, J., 1879, The Upper Silurian Bone Bed at the Rhymney River section: Cardiff Nat. Soc. Trans., v. 10, p. 92. (Southern Welsh Borderland)

Strachan, I., 1960, The Ordovician and Silurian graptolite zones in Britain: Internat. Geol. Cong., 21st, Copenhagen 1960, Rept., pt. 7, p. 109-113. (Correlation—General)

——1964, The Silurian Period: Geol. Soc. London Quart. Jour., v. 120, p. 237-240. (Correlation—General)

——1967, The geology of Wren's Nest, *in* Wren's Nest, National Nature Reserve: London, The Nature Conservancy, p. 9-12. (Northern Welsh Borderland)

——1971, A synoptic supplement to "A monograph of British graptolites by Miss G. L. Elles and Miss E.M.R. Wood": Palaeontographical Soc. Mon., 130 p. (Correlation—General)

Strahan, A., 1899, The geology of the South Wales coalfield. Pt. I. The country around Newport, Monmouthshire: Great Britain Geol. Survey Mem., sheet 249, new ser., 115 p. (Southern Welsh Borderland)

——1909, The geology of the South Wales coalfield. Pt. I. The country around Newport, Monmouthshire (2d ed.): Great Britain Geol. Survey Mem., sheet 249, 115 p. (Southern Welsh Borderland)

——1910, South Wales (Jubilee volume): Geol. Assoc. London, p. 826-858. (Southern Welsh Borderland)

Strahan, A., 1913a, Anniversary address of the president: Geol. Soc. London Proc., v. 69, p. 70-91. (Southeast England)

——1913b, Boring at the East Anglian Ice Company's works, Lowestoft: Great Britain Geol. Survey Summary Progress for 1912, p. 87-88. (Southeast England)

——1913c, Batsford (or Lower Lemington) boring, near Moreton-in-Marsh: Great Britain Geol. Survey Summary Progress for 1912, p. 90-91. (Southeast England)

——1916, On a deep boring for coal near Little Missenden, in Buckinghamshire: Great Britain Geol. Survey Summary Progress for 1915, p. 43-46. (Southeast England)

Strahan, A., and Cantrill, T. C., 1912, The geology of the South Wales coalfield. Pt. III. The country around Cardiff (2d ed.): Great Britain Geol. Survey Mem., sheet 263, new ser., 157 p. (Southern Welsh Borderland)

Strahan, A., and de Rance, C. E., 1890, The geology of the neighbourhoods of Flint, Mold, and Ruthin: Great Britain Geol. Survey Mem., sheet 79 S.E., old ser., 242 p. (North Wales)

Strahan, A., and Walker, A. O., 1879, On the occurrence of pebbles with upper Ludlow fossils in the lower Carboniferous Conglomerates of North Wales: Geol. Soc. London Quart. Jour., v. 35, p. 268-274. (North Wales)

Strahan, A., Tiddeman, R. H., Wilkinson, B.S.N., Cantrill, T. C., and Thomas, H. H., 1902, South Wales district—Glamorganshire and Carmarthenshire: Great Britain Geol. Survey Summary Progress for 1901, p. 34-53. (Central Wales)

Strahan, A., Cantrill, T. C., Dixon, E.E.L., and Thomas, H. H., 1907, The geology of the South Wales coalfield. Pt. VII. The country around Ammanford: Great Britain Geol. Survey Mem., 246 p. (Central Wales)

Strahan, A., Cantrill, T. C., Dixon, E.E.L., Thomas, H. H., and Jones, O. T., 1914, Geology of the South Wales coalfield. Pt. XI. The country around Haverfordwest: Great Britain Geol. Survey Mem., sheet 228, new ser., 262 p. (Southwest Wales)

Straw, S. H., 1929, The Siluro-Devonian boundary in south-central Wales: Manchester Geol. Assoc. Jour., v. 1, p. 79-102. (Central Wales)

——1933, The fauna of the Palaeozoic rocks of the Little Missenden boring: Great Britain Geol. Survey Summary Progress for 1932, Pt. II, p. 112-142. (Southeast England)

——1937, The higher Ludlovian rocks of the Builth district: Geol. Soc. London Quart. Jour., v. 93, p. 406-456. (Central Wales)

——1953, The Silurian succession at Cwm Graig Ddu (Breconshire): Geol. Jour., v. 1, p. 208-219. (Central Wales)

Straw, S. H., and Smith-Woodward, A., 1933, The fauna of the Palaeozoic rocks of the Little Missenden boring: Great Britain Geol. Survey Summary Progress for 1932, p. 112-142. (Southeast England)

Strickland, H. E., 1851, On the elevatory forces which raised the Malvern Hills: Philos. Mag., ser. 4, v. 2, p. 358-365. (Southern Welsh Borderland)

Stubblefield, C. J., 1939, Some Devonian and supposed Ordovician fossils from southwest Cornwall: Great Britain Geol. Survey Bull., no. 2, p. 63-71. (Southwest England)

Swanston, W., and Lapworth, C., 1878, Correlation of the Silurian rocks of County Down: Belfast Naturalists' Field Club Proc., new ser., v. 1, p. 107-148. (Ireland)

Sweeting, G. S., 1927, The petrography of the Malvern Quartzite, Hollybush Sandstone, and May Hill Sandstone exposed in the Eastnor (Herefordshire) district: Geol. Assoc. London Proc., v. 38, p. 548-560. (Southern Welsh Borderland)

Symes, R. G., and Bailey, W. H., 1879, Explanatory memoir to accompany sheets 41, 53, and 64 of the maps of the Geological Survey of Ireland, including the country around Ballina, Crossmolina, Killala, Foxford, and Ballycastle: Geol. Survey Ireland Mem., 38 p. (Ireland)

Symonds, W. S., 1872, Records of the rocks: London, John Murray, 433 p. (Southern Welsh Borderland)

——1884, Old stones (1st ed.): London, D. Bogue, 149 p. (Southern Welsh Borderland)

Symonds, W. S., and Lambert, A., 1861, On the sections of the Malvern and Ledbury Tunnels (Worcestershire and Herefordshire Railway) and the intervening line of railway: Geol. Soc. London Quart. Jour., v. 17, p. 152-160. (Southern Welsh Borderland)

Tarlo, L. B., 1965, Siluro-Devonian boundary: Geol. Mag., v. 102, p. 349-350. (Correlation—General)

Taylor, W.E.G., and Lister, T. R., 1964, A reappraisal of the revised Ludlow classification: Sheffield Univ. Geol. Soc. Jour., v. 5, p. 20-25. (Correlation—General)

Temple, J., 1968, The lower Llandovery (Silurian) brachiopods from Keisley, Westmorland: Palaeontographical Soc. Mon., 55 p. (Northern England)

——1970, The lower Llandovery brachiopods and trilobites from Ffridd Mathrafal, near Meifod, Montgomeryshire: Palaeontographical Soc. Mon., v. 124, 76 p. (Central Wales)

Termier, H., and Termier, G., 1964, Les temps fossilifères: I. Palaeozoique Inférieur (Vol. 1): Paris, Masson & Cie., 689 p. (British Isles—General)

Thomas, H. H., 1911, The Skomer Volcanic Series (Pembrokeshire): Geol. Soc. London Quart. Jour., v. 67, p. 175-214. (Southwest Wales)

Toghill, P., 1968a, The graptolite assemblages and zones of the Birkhill Shales (Lower Silurian) at Dobb's Linn: Palaeontology, v. 11, p. 654-658. (Southern Uplands)

——1968b, The stratigraphical relationships of the earliest Monograptidae, and the dimorphograptidae: Geol. Mag., v. 105, p. 46-51. (Correlation—General)

——1969, Discussion of the terms "Llandovery Series" and "Valentian Series" as time divisions of the Silurian: Geol. Soc. London Proc., no. 1651, p. 235-238. (Correlation—General)

——1970, The south-east limit of the Moffat Shales in the upper Ettrick Valley region, Selkirkshire: Scottish Jour. Geology, v. 6, p. 233-242. (Southern Uplands)

——1971, The distribution of the graptolite faunas across the Ordovician-Silurian boundary in Britain, in Babin, C., ed., Colloque ordovicien-silurien: Bur. Recherches Géol. Minières, Mém. no. 73, p. 417-422. (Correlation—General)

Toghill, P., and Strachan, I., 1970, The graptolite fauna of Grieston Quarry near Innerleithen, Pebbleshire: Palaeontology, v. 13, p. 511-521. (Southern Uplands)

Traill, W. A., and Egan, F. W., 1871, Explanatory memoir to accompany sheets 49, 50, and part of 61 of the maps of the Geological Survey of Ireland, including the country around Downpatrick, and the shores of Dundrum Bay and Strangford Lough; county of Down: Geol. Survey Ireland Mem., 71 p. (Ireland)

Tucker, E. V., 1960, Ludlovian biotite-bearing rocks: Geol. Mag., v. 97, p. 245-249. (Southern Welsh Borderland)

——1965, The Malvern line: Geol. Mag., v. 102, p. 88-90. (Southern Welsh Borderland)

Turner, J. S., 1935, Gotlandian vulcanicity in western Europe: Geol. Mag., v. 122, p. 145-151. (British Isles—General)

——1949, The deeper structure of central and northern England: Yorkshire Geol. Soc. Proc., v. 27, p. 280-297. (British Isles—General)

Turner, S., 1970, Timing of the Appalachian/Caledonian orogen contraction: Nature, v. 227, p. 90. (British Isles—General)

Tyler, W. H., 1925, Notes on sheet 48 N.W. (Shropshire): Geol. Assoc. London Proc., v. 36, p. 377-378. (Northern Welsh Borderland)

Van de Kamp, P. C., 1969, The Silurian volcanic rocks of the Mendip Hills, Somerset, and the Tortworth area, Gloucestershire, England: Geol. Mag., v. 106, p. 542-553. (Southern Welsh Borderland)

Vernon, R. D., 1912, On the geology and palaeontology of the Warwickshire coalfield: Geol. Soc. London Quart. Jour., v. 68, p. 587-638. (Northern Welsh Borderland)

Vine, G. R., 1887, Notes on the palaeontology of the Wenlock Shales of Shropshire (Mr. Maw's washings, 1880): Yorkshire Geol. Soc. Proc., v. 9, p. 225-248. (Northern Welsh Borderland)

Wade, A., 1911, The Llandovery and associated rocks of northeastern Montgomeryshire: Geol. Soc. London Quart. Jour., v. 67, p. 415-459. (Central Wales)

Walder, P. S., 1941, The petrography, origin, and conditions of deposition of a sandstone of Downtonian age: Geol. Assoc. London Proc., v. 52, p. 245-256. (Wales and Welsh Borderland—General)

Walliser, O. H., 1962, Conodontenchronologie des Silurs (Gotlandium) und des tieferen Devons mit besonderer Berücksichtigung der Formationsgrenze: Symposium Silur/Devon-Grenze, Bonn-Bruxelles 1960, p. 281-287.

——1964, Conodonten des Silurs: Hess, Landesamt Bodenforschung Abh., v. 41, 106 p.

Walmsley, V. G., 1959, The geology of the Usk inlier (Monmouthshire): Geol. Soc. London Quart. Jour., v. 114, p. 483-521. (Southern Welsh Borderland)

——1962, Upper Silurian-Devonian contacts in the Welsh Borderland and South Wales: Symposiums-Band der 2. internationalen Arbeitstagung über die Silur Devon-Grenze und die Stratigraphie von Silur und Devon: Stuttgart, E. Schweizerbart'sche Verlags., p. 287-295. (Wales and Welsh Borderland—General)

Walton, E. K., 1955, Silurian greywackes in Peebleshire: Royal Soc. Edinburgh Proc., ser. B, v. 65, p. 327-357. (Southern Uplands)

——1956, Limitations of graded bedding, and alternative criteria of upward sequence in the rocks of the Southern Uplands: Edinburgh Geol. Soc. Trans., v. 16, p. 262-271. (Southern Uplands)

——1960, Some aspects of the succession and structure in the lower Palaeozoic rocks of the Southern Uplands of Scotland: Geol. Rundschau, v. 50, p. 63-77. (Southern Uplands)

——1963, Sedimentation and structure in the Southern Uplands, in Johnson, M.R.W., and Stewart, F. H., eds., The British Caledonides: Edinburgh and London, Oliver & Boyd, p. 71-97. (Southern Uplands)

——1965, Lower Palaeozoic rocks, in Craig, G. Y., ed., The geology of Scotland: Hamden, Conn., Archon Books, p. 161-227. (Southern Uplands)

——1968, Some rare sedimentary structures in the Silurian rocks of Kirkcudbrightshire: Scottish Jour. Geology, v. 4, p. 355-369. (Southern Uplands)

Ward, J. C., 1879, On the physical history of the English Lake District, with notes on the possible subdivisions of the Skiddaw Slates: Geol. Mag., dec. 2, v. 6, p. 49-61 and 110-125. (Northern England)

Warren, P. T., 1963, The petrography, sedimentation, and provenance of the Wenlock rocks near Hawick, Roxburghshire: Edinburgh Geol. Soc. Trans., v. 19, p. 225-255. (Southern Uplands)

——1964, The stratigraphy and structure of the Silurian (Wenlock) rocks south-east of Hawick, Roxburghshire, Scotland: Geol. Soc. London Quart. Jour., v. 120, p. 193-222. (Southern Uplands)

——1970, Llanrwst, North Wales, 19th to 22nd September, 1969: Yorkshire Geol. Soc. Proc., v. 38, p. 68-74. (North Wales)

——1971, The sequence and correlation of graptolite faunas from the Wenlock-Ludlow rocks of North Wales, in Babin, C., ed., Colloque ordovicien-silurien: Bur. Recherches Géol. et Minières Mém., no. 73, p. 451-460. (North Wales)

Warren, P. T., Rickards, R. B., and Holland, C. H., 1966, Pristigraptus ludensis (Murchison 1839)—Its synonymy and allied species—And the position of the Wenlock/Ludlow boundary in the Silurian graptolite sequence: Geol. Mag., v. 103, p. 466-467. (Correlation—General)

Warren, P. T., Harrison, R. K., Wilson, H. E., Smith, E. G., and Nutt, J. C., 1970, Tectonic ripples and associated minor structures in the Silurian rocks of Denbighshire, North Wales: Geol. Mag., v. 107, p. 51-60. (North Wales)

Waterschoot van der Gracht, W.A.J.M. van, 1938, A structural outline of the Variscan front and its foreland from south-central England to eastern Westphalia and Hessen: Compte Rendu II, Cong. Strat. Carbonifère, Heerlen, v. 3, p. 1485-1565. (Southeast England)

Waterston, C. D., 1965, Old Red Sandstone, in Craig, G. Y., ed., The geology of Scotland: Hamden, Conn., Archon Books, p. 269-308. (Midland Valley)

Watney, G. R., and Welch, E. G., 1910, The graptolite zones of the Salopian rocks of the Cautley area near Sedbergh, Yorkshire: Geol. Mag., dec. 5, v. 7, p. 473. (Northern England)

——1911, The zonal classification of the Salopian rocks of Cautley and Ravenstonedale: Geol. Soc. London Quart. Jour., v. 67, p. 215-237. (Northern England)

Watts, W. W., 1891, The geology of the Long Mountain, on the Welsh Borders: Rept. British Assoc. Adv. Sci., Leeds, 1890, Trans. Secs., p. 817. (Northern Welsh Borderland)

——1925, The geology of South Shropshire: Geol. Assoc. London Proc., v. 36, p. 322-363. (Northern Welsh Borderland)

Wedd, C. B., 1932, The principles of Palaeozoic and later tectonic structure between the Longmynd and the Berwyns: Great Britain Geol. Survey Summary Progress for 1931, pt. 2, p. 1-22. (Northern Welsh Borderland)

Wedd, C. B., and King, W.B.R., 1924, The geology of the country around Flint, Hawarden, and Caergwrle: Great Britain Geol. Survey Mem., sheet 108, new ser., 222 p. (North Wales)

Wedd, C. B., Smith, B., and Wills, L. J., 1927, The geology of the country around Wrexham. Pt. I: Great Britain Geol. Survey Mem., sheet 121, new ser., 179 p. (North Wales)

Wedd, C. B., Smith, B., King, W.B.R., and Wray, D. A., 1929, The country around Oswestry: Great Britain Geol. Survey Mem., sheet 137, new ser., 234 p. (North Wales)

Weir, J. A., 1960, Mudstone inclusions in Salopian conglomerates from County Clare: Geol. Mag., v. 97, p. 283-288. (Ireland)

——1962, Geology of the lower Palaeozoic inliers of Slieve Bernagh and the Cratloe Hills, County Clare: Royal Dublin Soc. Sci. Proc., ser. A, v. 1, p. 233-263. (Ireland)

——1968, Structural history of the Silurian rocks of the coast west of Gatehouse, Kirkcudbrightshire: Scottish Jour. Geology, v. 4, p. 31-52. (Southern Uplands)

Weir, J. A., 1973, Lower Palaeozoic graptolitic facies in Ireland and Scotland: Review, correlation, and palaeogeography: Royal Dublin Soc. Sci. Proc., ser. A, v. 4, p. 439-460. (British Isles—General)

Welch, F.B.A., 1962, South-west England and south Wales district: Great Britain Geol. Survey Summary Progress for 1961, p. 28-29. (Southern Welsh Borderland)

Wells, A. K., and Kirkaldy, J. F., 1966, Outline of historical geology (5th ed.): London, Murby, p. 503. (British Isles—General)

Westoll, T. S., 1945, A new cephalaspid fish from the Downtonian of Scotland, with notes on the structure and classification of ostracoderms: Royal Soc. Edinburgh, v. 61, p. 341-357. (Midland Valley)

——1951, The vertebrate-bearing strata of Scotland: Internat. Geol. Cong., 18th, London 1948, Rept., pt. 11, p. 5-21. (Midland Valley)

Wethered, E., 1893, On the microscopic structure of the Wenlock Limestone: Geol. Soc. London Quart. Jour., v. 49, p. 236-248. (Northern Welsh Borderland)

Whitaker, J.H.McD., 1962, The geology of the area around Leintwardine: Geol. Soc. London Quart. Jour., v. 118, p. 319-351. (Northern Welsh Borderland)

Whitaker, W., 1877, The geology of the eastern end of Essex (Walton Naze and Harwich): Great Britain Geol. Survey Mem., sheet 48, S.E., old ser., 32 p. (Southeast England)

——1906, Water supply of Suffolk from underground sources: Great Britain Geol. Survey Mem., 177 p. (Southeast England)

——1908a, Water supply of Kent: Great Britain Geol. Survey Mem., 399 p. (Southeast England)

——1908b, On the finding of Silurian beds in Kent: Geol. Mag., dec. 5, v. 5, p. 469. (Southeast England)

——1909, On the finding of Silurian beds in Kent: Rept. British Assoc. Adv. Sci., Dublin, 1908, Trans. Secs., p. 711. (Southeast England)

——1921, Water supply of Buckinghamshire and Herefordshire: Great Britain Geol. Survey Mem., 368 p. (Southeast England)

Whitaker, W., and Jukes-Browne, A. J., 1894, On deep borings at Culford and Winkfield, with notes on those at Ware and Chestnut: Geol. Soc. London Quart. Jour., v. 50, p. 488-514. (Southeast England)

Whitaker, W., and Thresh, J. C., 1916, Water supply of Essex from underground sources: Great Britain Geol. Survey Mem., 510 p. (Southeast England)

White, E. I., 1950, The vertebrate faunas of the lower Old Red Sandstone of the Welsh Borders: British Mus. (Nat. History) Bull., Geology, v. 1, p. 51-67, Pl. 5. (Wales and Welsh Borderland—General)

Whitehead, T. H., 1962, Recent geological work in Shropshire and neighbouring counties: Shrewsbury, Caradoc, and Severn Valley Field Club Trans., v. 15, p. 1-18. (Northern Welsh Borderland)

Whitehead, T. H., and Eastwood, T., 1927, The geology of the southern part of the South Staffordshire coalfield: Great Britain Geol. Survey Mem., 218 p. (Northern Welsh Borderland)

Whitehead, T. H., and Pocock, R. W., 1947, Dudley and Bridgnorth: Great Britain Geol. Survey Mem., sheet 167, new ser., 226 p. (Northern Welsh Borderland)

Whittard, W. F., 1925, Notes on the Valentian rocks of Shropshire: Geol. Assoc. London Proc., v. 36, p. 378-381. (Northern Welsh Borderland)

——1928, The stratigraphy of the Valentian rocks of Shropshire—The main outcrop: Geol. Soc. London Quart. Jour., v. 83, p. 737-759. (Northern Welsh Borderland)

——1931a, The geology of the Ordovician and Valentian rocks of the Shelve country, Shropshire: Geol. Assoc. London Proc., v. 42, p. 322-339. (Northern Welsh Borderland)

——1931b, Easter field meeting (extension) to Minsterley, 8-11 April 1931: Geol. Assoc. London Proc., v. 42, p. 339-344. (Northern Welsh Borderland)

——1932, The stratigraphy of the Valentian rocks of Shropshire—The Longmynd-Shelve and Breidden outcrops: Geol. Soc. London Quart. Jour., v. 88, p. 859-902. (Northern Welsh Borderland)

——1952, A geology of South Shropshire: Geol. Assoc. London Proc., v. 63, p. 143-197. (Northern Welsh Borderland)

——1953, Report of summer field meeting in South Shropshire, 1952: Geol. Assoc. London Proc., v. 64, p. 232-250. (Northern Welsh Borderland)

——1961, Silurian: Lexique stratigraphique International, Fasc. 3aV. Europe: Paris, Centre Natl. Recherche Sci., 273 p. (British Isles—General)

Whittard, W. F., and Smith, S., 1944, Unrecorded inliers of Silurian rocks near Wickwar, Gloucestershire, with notes on the occurrence of a stromatolite: Geol. Mag., v. 81, p. 65-76. (Southern Welsh Borderland)

Whittington, H. B., 1938, The geology of the district around Llansantffraid ym Mechain, Montgomeryshire: Geol. Soc. London Quart. Jour., v. 94, p. 423-457. (Central Wales)

Whitworth, T., 1952, Malvern tectonics: Geol. Mag., v. 89, p. 384. (Southern Welsh Borderland)

——1962, Malvern structures: Geol. Mag., v. 99, p. 375-377. (Southern Welsh Borderland)

Wilkinson, S. B., and Kilroe, J. R., 1881, Explanatory memoir to accompany sheet 57 of the maps of the Geological Survey of Ireland, including parts of Fermanagh, Monaghan, and Cavan: Geol. Survey Ireland Mem., 22 p. (Ireland)

Wilkinson, S. B., Kilroe, J. R., and Bailey, W. H., 1882, Explanatory memoir to accompany sheet 45 of the maps of the Geological Survey of Ireland, including the country around Enniskillen, Fivemiletown, Trillick, Lisbellaw, and Maguiresbridge in the counties of Fermanagh and Tyrone: Geol. Survey Ireland Mem., 23 p. (Ireland)

Williams, A., 1951, Llandovery brachiopods from Wales with special reference to the Llandovery district: Geol. Soc. London Quart. Jour., v. 107, p. 85-136. (Correlation—General)

——1953, The geology of the Llandeilo district, Carmarthenshire: Geol. Soc. London Quart. Jour., v. 108, p. 177-207. (Central Wales)

Williams, B. J., and Prentice, J. E., 1958, Slump structures in the Ludlovian rocks of North Herefordshire: Geol. Assoc. London Proc., v. 68, p. 286-293. (Northern Welsh Borderland)

Williams, G. J., 1907, Note on the geological age of the shales of the Parys Mountain, Anglesey: Geol. Mag., dec. 5, v. 4, p. 148-150. (North Wales)

Wills, L. J., 1920, The geology of the Llangollen district: Geol. Assoc. London Proc., v. 31, p. 1-15. (North Wales)

——1929, The physiographical evolution of Britain: London, Edward Arnold & Co., 375 p. (British Isles—General)

——1935, An outline of the palaeogeography of the Birmingham country: Geol. Assoc. London Proc., v. 46, p. 211-246. (Northern Welsh Borderland)

——1948, The palaeogeography of the Midlands: Liverpool, Univ. Press, 147 p. (Northern Welsh Borderland)

——1951, Palaeogeographic atlas of the British Isles and adjacent parts of Europe: London, Blackie, 64 p. (British Isles—General)

Wills, L. J., and Laurie, W. H., 1938, Deep sewer trench along the Bristol road from Ashill road near the Longbridge Hotel to the city boundary at Rubery, 1937: Birmingham Nat. History Philos. Soc. Proc., v. 16, p. 175-180. (Northern Welsh Borderland)

Wills, L. J., and Smith, B., 1922, The lower Palaeozoic rocks of the Llangollen district, with special reference to the tectonics: Geol. Soc. London Quart. Jour., v. 78, p. 176-226. (North Wales)

Wills, L. J., Wilkens, T. G., and Hubbard, G. H., 1925, The upper Llandovery series of Rubery: Birmingham Nat. History Philos. Soc. Proc., v. 15, p. 67-83. (Northern Welsh Borderland)

Wilson, H. E., 1965, North Wales district, Rhyl (95) sheet, Denbigh (107) sheet: Great Britain Geol. Survey Summary Progress for 1964, p. 60-62. (North Wales)

——1966, North Wales district, Rhyl (95) sheet, Denbigh (107) sheet: Great Britain Geol. Survey Summary Progress for 1965, p. 57-58. (North Wales)

Wilson, V., 1962, Midlands and Central and North Wales: Great Britain Geol. Survey Summary Progress for 1961, p. 34-37. (Northern Welsh Borderland)

Wood, A., 1961, The lower Palaeozoic turbidites of Wales: Geol. Soc. London Proc., no. 1587, p. 74. (Central Wales)

Wood, A., and Smith, A. J., 1959, The sediments and sedimentary history of the Aberystwyth Grits (upper Llandoverian): Geol. Soc. London Quart. Jour., v. 114, p. 163-195. (Central Wales)

Wood, E.M.R., 1900, The Lower Ludlow Formation and its graptolite fauna: Geol. Soc. London Quart. Jour., v. 56, p. 415-492. (Wales and Welsh Borderland—General)

——1904, The graptolites of the Lower Ludlow Shales: Geol. Assoc. London Proc., v. 18, p. 446-451. (Wales and Welsh Borderland—General)

Wood, E.M.R., 1906, On the Tarannon Series of Tarannon: Geol. Soc. London Quart. Jour.,
    v. 62, p. 644-701. (Central Wales)

Woods, E. G., and Crosfield, C., 1925, The Silurian rocks of the central part of the Clwydian
    Range: Geol. Soc. London Quart. Jour., v. 81, p. 170-194. (Northern Wales)

Woodward, A. S., 1904, Notes on the geology and fossils of the Ludlow district: Geol.
    Assoc. London Proc., v. 18, p. 429-442. (Northern Welsh Borderland)

Woodward, H. B., 1887, The geology of England and Wales, with notes on the physical
    features of the country (2d ed.): London, G. Philip & Son. (British Isles—General)

Wooldridge, S. W., and Linton, D. L., 1938, Some episodes in the structural evolution
    of S.E. England considered in relation to the concealed boundary of Meso-Europe:
    Geol. Assoc. London Proc., v. 49, p. 264-291. (Southeast England)

Wright, A. D., 1967, A note on the stratigraphy of the Kildare inlier: Irish Naturalists'
    Jour., v. 15, p. 340-343. (Ireland)

Wright, J. E., 1968, The geology of the Church Stretton area (explanation of 1:25,000 geological
    sheet SO49): Great Britain Geol. Survey, 89 p. (Northern Welsh Borderland)

Wright, J. R., 1832, On the secondary formations in the neighbourhood of Ludlow: Geol.
    Soc. London Proc., v. 1, p. 387-388. (Northern Welsh Borderland)

Ziegler, A. M., 1964, The Malvern line: Geol. Mag., v. 101, p. 467-469. (Southern Welsh
    Borderland)

——1965, Silurian marine communities and their environmental significance: Nature, v. 207,
    p. 270-272. (Wales and Welsh Borderland—General)

——1966, The Silurian brachiopod, *Eocoelia hemisphaerica* (J. de C. Sowerby), and related
    species: Palaeontology, v. 9, p. 523-543. (Correlation—General)

——1970, Geosynclinal development of the British Isles during the Silurian Period: Jour.
    Geology, v. 78, no. 4, p. 445-479. (British Isles—General)

——1972, Some points of interest concerning the Silurian inliers of southwest central Ireland
    in their geosynclinal context: A reply: Geol. Soc. London Jour., v. 128, p. 263-266.
    (Ireland)

Ziegler, A. M., Boucot, A. J., and Sheldon, R. P., 1966, Silurian pentameroid brachiopods
    preserved in position of growth: Jour. Paleontology, v. 40, p. 1032-1036. (Midland Valley)

Ziegler, A. M., Cocks, L.R.M., and Bambach, R. K., 1968a, The composition and structure
    of Lower Silurian marine communities: Lethaia, v. 1, p. 1-27. (Wales and Welsh
    Borderland—General)

Ziegler, A. M., Cocks, L.R.M., and McKerrow, W. S., 1968b, The Llandovery transgression
    of the Welsh Borderland: Palaeontology, v. 11, p. 736-782. (Wales and Welsh Border-
    land—General)

Ziegler, A., McKerrow, W. S., Burne, R. V., and Baker, P. E., 1969, The age and environmental
    setting of the Skomer Volcanic Group, Pembrokeshire: Geol. Assoc. London Proc.,
    v. 80, p. 409-439. (Southwest Wales)

MANUSCRIPT RECEIVED BY THE SOCIETY AUGUST 30, 1973